BROKEN GENIUS

BROKEN GENIUS

The rise and fall of William Shockley, creator of the electronic age

Joel N. Shurkin

Macmillan

London   New York   Melbourne   Hong Kong

First published in hardback 2006
First published in paperback 2008 by
Macmillan
Houndmills, Basingstoke, Hampshire RG21 6XS and
175 Fifth Avenue, New York, N. Y. 10010
Companies and representatives throughout the world

ISBN-13: 978–1–4039–8815–7 hardback
ISBN-10: 1–4039–8815–3 hardback

ISBN-13: 978–0–230–55192–3 paperback
ISBN-10: 0–230–55192–0 paperback

This book is printed on paper suitable for recycling and made from fully managed and
sustained forest sources. Logging, pulping and manufacturing processes are expected
to conform to the environmental regulations of the country of origin.

A catalogue record for this book is available from the British Library.

A catalog record for this book is available from the Library of Congress.

Transferred to Digital Printing 2014

Contents

For Carol
Ani L'Dodi, v'Dodi Li

Preface

I believe that William Shockley was, in terms of practical impact on the world, one of the most important scientists of the twentieth century. *Time* magazine agreed in its *fin de siècle* issue. He led the group at Bell Telephone Laboratories that created the seminal invention of the modern world, the transistor. The transistor begot the integrated circuit; the integrated circuit begot the microprocessor; and when their device was combined with the computer (invented only a few years earlier), the greatest social and economic upheaval since the Industrial Revolution almost two centuries earlier was inevitable. Every home in the developed world has thousands or even millions of transistors. World commerce totally depends on them, as do health care, culture, defense, transportation, increasingly art – and civilization in general.

Forty years ago, while researching a book in sub-Saharan Africa, I watched Somali nomads take transistor radios out of their camel-skin bags to listen to newscasts. Somalia and Ethiopia then were at war over a patch of the desert and the nomads wanted to make sure they kept their flocks from stumbling into the shooting. It was their only modern artifact – except for the Coke bottles they used to plug the holes in the peaks of their skin tents.

Along with his Bell Labs colleagues, Walter Brattain and John Bardeen, William Shockley won the Nobel Prize for Physics in 1956.

He was a key player in the development of the modern science of operations research. During the Second World War, he used statistics to show the Air Corps how to maximize its bombing efficiency and the Navy how to destroy more U-boats and improve the use of airborne radar. His work also may have saved thousands of lives in the North Atlantic convoys to Britain. He set up training programs for both services. By the end of the war in the Pacific, American bombers, drilled by Shockley's method, were hitting 95% of their targets through the clouds. He may even have played a minor role in the decision to drop the atomic bomb on Japan, marshaling the numbers to demonstrate that

B-29 bombing alone would not do enough damage to lessen the blood-bath of an Allied invasion. His efforts won the National Medal of Merit, the highest possible civilian decoration.

He was a member of the National Academy of Sciences, the most prestigious scientific body in America, elected by his peers.

He put the 'silicon' in Silicon Valley, and his failed company was the grandfather of all Silicon Valley companies, although it ended up breaking his heart.

He was a tenured professor at Stanford University, was happily married and had all the money he needed to live happily, quietly and well. He chose not to. He instead set himself up for public ridicule and squandered his public reputation.

Many scientists have stretched beyond their field and made fools of themselves, or harmed their good name. Few have gone to the lengths Bill Shockley did to earn the opprobrium of his peers or the public. He quickly lost all of his friends; indeed, his oldest friend became his most potent enemy. He became notorious, a scientific pariah.

I wanted to know how that could happen.

So discredited did Bill Shockley become that a kind of revisionist history took over. 'Oh, you know he really wasn't that important to the invention of the transistor,' I was told confidentially dozens of times, usually with a satisfied smile. 'The other two actually invented the transistor.' Wink, wink, nudge, nudge. The destruction of his good name was almost total, and as you will see, almost inevitable. Shockley was impelled toward self-destruction.

That is the story this book chronicles. Sophocles could have written the plot.

Uncovering such a complex saga seemed daunting at first. Fortunately, Shockley's family shared one strange quirk. They never threw anything out. Several rooms and the garage at the Shockley home on the Stanford campus were stuffed with documents, letters, folders, computer, video and audiotapes, notebooks, diaries, memos, and files. We had to crack open two safes to get at all the material. And that didn't include the dozens of boxes already donated to the Stanford University archives, more than 60 linear feet of stuff, unimaginable stuff. The Shockley family was in some ways both a biographer's delight and worst nightmare.

Hidden in all those documents and artifacts and in the recollections of those who knew him was, I hoped, the answer to the enigma that

William Shockley posed: Why would a man as unquestionably brilliant as he knowingly and deliberately destroy himself?

Joel Shurkin
January 2006

PART I
Moira: May and Billy

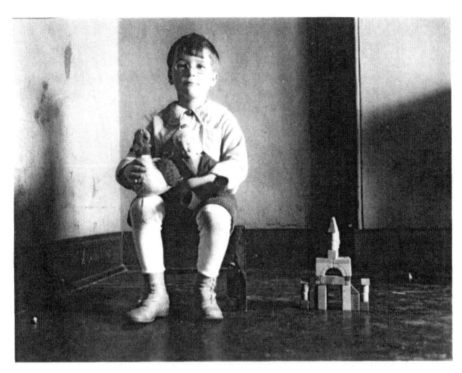

Figure 1 William Shockley, aged 5 (1915).

CHAPTER 1
'I've got dark eyes. I can frighten people.'

William Bradford Shockley was born on 13 February 1910 to an eccentric American couple living in London. May Bradford Shockley and her husband William grew grateful that he was an only child.

May would live all but 14 years of her son's life, seeing him rise to become one of the most famous scientists in the world and, later, one of the most vilified, sharing his triumphs and ignoring his failures – the self-proclaimed 'grandmother of the transistor.' William would miss it all.

May Shockley grew up in New Mexico and Missouri in the home of her mother and stepfather; a slightly asthmatic girl with a mind of her own. She was capable of packing a rifle and riding out onto the desert or grasslands on horseback to pop rabbits, or, when the mood struck her, to capture them with traps and bash their heads in on rocks. Something of a math protégé with an artistic bent, she found her way to Stanford University in Palo Alto, California, mostly because it was coed and free. She went on geology field trips and took up rock climbing, and is believed to be the first person to climb Mount Whitney solo. She picked up enough geology to ride by stagecoach to the roaring mining town of Tonopah, Nevada, to help her stepfather's surveying business, becoming the first female US Deputy Mineral Surveyor.

May was undaunted in one of the last towns of the Old West where women were not plentiful. She was a splendid shot – and totally uninterested in the men around her. She was not especially pretty – slim with an oval face, wide-set dark blue eyes, a sharp triangular nose and wide mouth. Her most striking feature was her dark blonde hair, which reached down almost to the back of her knees. She wore it up and seemed determined that no man would see it down – until she met William Shockley senior, an MIT-trained mining engineer.

Figure 2 May Bradford Shockley, 1931.

Shockley *père* spoke eight languages and speculated in mines for a living, although he was a better linguist than a businessman. He had traveled around the world in a life of emprise and danger, and when May met him she wrote back to her mother: 'I was amazed to find someone in the middle of Nevada who could talk to me about Italian paintings.' At 52, he was 22 years older than she, with a goatee about to become pure white. He came from one of America's most illustrious families, the direct descendant of John Alden and Priscilla Mullins of *Mayflower* fame. His father was one of the last whaling captains to sail from New Bedford, and his maternal grandfather built the ships.

Several of William senior's ancestors had a hand in founding the Massachusetts Institute of Technology, so he went there, seemingly because he had nothing better to do. A mediocre student in mathematics, he had trouble finding work but wound up with mining companies in California and Nevada. He studied music in New York and languages in Europe, and finally found jobs with mining firms in London in the last days of foreign mineral concessions. He roamed all over Asia, often shooting his way out of jams. There even was a price on his head in one Chinese province. His collection of porcelain would eventually find its way into the Stanford University Museum.

Small wonder that May, in the wastelands of Tonopah, was swept off her feet. William was everything the outside world promised and not like any man she had met before. His age was to his benefit, she felt; his maturity and wisdom were beyond anything in her experience. They married on 20 January 1908 and sailed for London.

How May felt about William is never clear in either of their diaries and letters. She married him eagerly, obviously respected him and may have loved him, although there is no remaining record of her actually having said so, and there are hints that she had an affair toward the end of her husband's life. He clearly adored her. Letters from his trips and expeditions were full of longing and loneliness, and sometimes, when they quarreled, he was clearly upset and remorseful. None of his letters to her contained the words 'I love you.' They appeared happily married.

Every night he would sit down at a typewriter and chronicle the day's events in sometimes painfully intimate detail. Some pages he duplicated and sent as letters home, but most he kept in green or gray books, some pages marked confidential. May read them and sometimes made corrections or additions. His diaries were so intimate and so complete that May and their son were able to imagine whole days lived more than a half century before.

This meticulous notation of his life set a theme in the years that followed. Just as he noted and saved everything he came across – it is possible to reconstruct meals eaten at restaurants on no particular occasion – his wife found it impossible to throw anything away. Their son would later acquire this same obsession.

The Shockleys led a gay life in London, limited only by William's inability to make money. There were several other MIT and Stanford engineers in the city the Shockleys were intimate with, including Herbert Hoover and his wife Lee Henry. Hoover later became president of Stanford, as well as, of course, the United States.

Strapped for cash, the family moved from flat to flat. They were demanding of their landlords and their servants – when they could afford them – and ferocious in defending their privacy, and left behind considerable ill will. Their inability to settle down in one place for more than a few months would be a theme in their lives for years.

May's morning sickness began on 12 June 1909, but neither she nor her husband understood what was happening. If May knew little about these things, she was still ahead of her husband.[2] He had not had that much to do with the pregnancy; a job opened in Siberia and he spent much of the time traveling up and down the Amur river, corresponding in code to save money and protect their privacy from telegraph operators. Mrs Hoover helpfully sent a book listing all the things that can go wrong with a pregnancy, and after reading it May announced she was never going to get pregnant again.

They were now living at 69 Victoria Street near Westminster, a flat they chose after May turned down 40 apartments. It cost £100 for six months, more than they could afford. They hired Sarah E. Richmond to be family nurse. She had been head of all the British army nurses, which would prove to have been good training.

At 2:30 in the morning of the 13th of February, May went into labor. At 9:45 the next morning, William recorded in his diary that he could hear the sound of his son crying, a 'strong penetrating lusty cry.' No doubt. May, who on first sight of the doctor had demanded chloroform, witnessed none of it. Most women probably swear sometime during labor that they will never go through it again, but May meant it. She had been in labor for more than 24 hours and enjoyed none of the experience

Figure 3 May and little Bill, London, 1910.

or the pregnancy that preceded it. William agreed completely. He was so shocked at the gore and pain he vowed that he would never subject his wife to it again. 'For a long time I could take no interest in the baby for my nerves were much shaken with May's agony. It is certainly a damnable business.... I'm glad it's over.'[2]

They named the child William Bradford Shockley.

William Shockley the elder, trained as an engineer, viewed his son's development as if he were watching a construction project. He was fascinated, confounded, and not entirely sure of how children normally develop and what they need and do. For a while, it seemed a grand experiment.

Everything wound up in father Shockley's diaries, sometimes in painful detail. Under his analytical veneer shimmered the light of amazement. Nothing escaped the chronicler's eye. We know that Bill was circumcised on 21 March; that 'Billy discovered that it gave him a pleasant sensation to rub powder puff between his legs after the bath' on the 13th of May; and that Bill's first erection happened in early December that year.[2] We know what the child weighed every day of his infancy. At five months, the boy could call himself 'Billy.' At a year, he could count to four and could tell if one of six objects was missing. He knew A, B, C, I, O, S. His pronunciation was good. 'His intelligence is developing quite rapidly,' William wrote, impressed. May was too: 'Billy is going ahead mentally very fast and is very well. We are very proud of him; he is no world-beater and shows no signs whatever of being anything more than a bright little boy.'[3]

May, meanwhile, suffered from acute post-partum depression, sometimes gaining solace by writing poetry, verses wracked with sadness and lonesomeness. She showed no one the poems. Her son found them in her papers after she died.

$\prec$

Two important aspects of Bill Shockley's early life become clear through his father's extensive and sometimes excruciatingly frank narratives. First: its instability. Nothing in the diaries indicated discord between May and William or that they ever treated each other with disrespect. They appear to have loved their son, doting even, although they seemed unable to show affection openly to him or each other. The instability came from their financial condition, their lifestyle, and the fact they were strangers in another country and had more trouble adapting than they perhaps realized. And they shared more than a tinge of paranoia.

Their Victoria Street flat was cramped. Bill had a pen in the nursery where he could crawl on a wool mat with coarse embroidered patterns and he could play in the garden, but the flat still was small. Two months after Bill was born, they moved to 5 Abington Court in Kensington and then 2 Camden House Terrace. Meanwhile, their financial situation got worse.

William tried to sell some of his stock in mines and rubber plantations to increase his income, but found no market. The only job offer he had was a vague chance to go to West Africa, which he turned down; he couldn't leave May and the baby. He sent off 491 letters to engineers, mostly in the US and Mexico, looking for investments. 'This will make me a lot of work,' he wrote, 'with one chance of making money in 10,000.' He sent off 270 letters in one day in August 1910. For all that effort, he got 30 letters in response and found a few likely prospects, none of which apparently panned out.[4] He kept trying and once had as many as 1,000 letters in circulation and 'no profit in sight.' He wanted to buy a house in London – they sold for about $20,000 – but it would cost him $8,000 a year to keep it up and he did not have that money. He had property in the San Francisco area, but couldn't sell that at a fair price.

Little Bill also had to cope with a succession of servants. The Shockleys had an elemental social problem with British help and did not trust strangers. The estimable nurse Richmond came and went several times, sometimes because William couldn't pay her. Her absence greatly upset the boy. Once William offered to pay her expenses on one of their tours of Europe in lieu of her salary, which he didn't have, and she accepted. May needed the help. She could not manage without someone to watch Bill as her post-partum depression lingered.

May fought constantly with cooks and housekeepers. William and May fought with landlords. Servants gossiped about the family to neighbors' servants and to property owners, which upset the Shockleys further. Their sense of privacy and security seemed perpetually threatened.

When they had to leave a flat, they crossed the Channel, sometimes with nurse in tow, touring the Continent at least three times a year. Then they returned to London and set up camp in a hotel while May scoured the city for an acceptable flat. Sometimes that took weeks. Once they found one they moved in, and Bill found himself in yet another bedroom with different servants. They ate out constantly; May had no talent for or interest in cooking. Every time they went out, Bill was left with a nurse.

The other striking aspect of Bill's childhood is his violent temper. Almost from the first day, Bill Shockley showed explosions of anger, one that amazed, disturbed and eventually cowed his parents. The outbursts got worse as time went on. Within a month of his birth, his father could write: '[He] gives signs of having a violent temper, and will very likely prove a difficult subject to handle. He is a good baby in keeping still when shown to visitors.'

Bill's temper grew to a constant presence – the bad tantrums that would try even the most even-tempered and experienced parents, which could hardly describe the Shockleys. All children throw tantrums, particularly around the age of four. But Bill's temper was extreme, going well beyond what most would consider the norm.

By eight months, Bill was biting people on the cheek with all of his four teeth, 'but only in play,' his father assures himself. Before he reached a year old, William senior wrote: 'He has a violent enough temper and when he was to eat, howls uninterruptedly at the top of his voice.'[2] Just after Bill's first birthday, William noted more violent temper tantrums: Bill 'screaming at the top of his voice and bending and throwing himself back until it seemed as if he [would] break his neck or hurt his head by hitting something.' His biting became nastier. Bill 'has bitten his mother severely many times and has slapped both his father and mother too often to record,' William wrote on June 24. 'It is an odd day when he does not break something,' William lamented.[2]

Mealtimes were the worst. One afternoon May tied Bill to his chair with a cloth napkin to keep him from hurling himself across the table as he screamed and threw beans on the floor. He could be dangerous to himself and others. Once he fell off his chair in rage, hitting his head on an iron radiator. He developed a habit of twisting his finger in his hair and one night he twisted it so tightly the hair had to be cut to extricate his finger. One day he threw a stone and hit a dachshund between the eyes.

Most of the tantrums were kept for his parents – he was active and charming when others were around – and neither William nor May had any idea how to handle him.

They had three choices: beat him into better manners, use psychological ploys to modify his behavior, or surrender. 'I should probably have beaten him had it not been for May,' William wrote. 'She would not allow it, and I think she is right, for [tho] it is easy enough to spank or shake a child it is most difficult to do it without becoming angry, and there is danger of permanently hurting a child or of losing his affection. Billy always gets angry because he is thwarted or denied something...'[5]

A week later: 'Billy was spanked today for the first time; he screamed and would not stop, so May spanked him, but he still kept screaming. It surprised him a good deal but it did not worry him after he was spanked. When he went to bed he would not keep still and May went to him, waving her hands and telling him she was a bad mother and not Billy's nice mother at all, and if he would be a good boy his good mother would come back.'[24]

Corporal punishment failed, so without any other obvious peaceful solution, they tried the third alternative: surrender. To avoid incidents and tantrums, the Shockleys decided to mollify Bill in any way they could. The situation was so bad that at least two servants, including Richmond, the former army nurse, quit. They hired a new nurse but Bill was unhappy. 'I believe he noticed the new nurse,' May wrote. 'I thought I had a good one but I cannot endure her and it breaks my heart to see her touch the baby.... If Miss Richmond would only come back.'[7] She would not. The inability of the Shockleys to control their son and their refusal to support her discipline was an impossible situation.

When they lost yet another apartment, they gave up on London and returned to the US, arriving in April 1913 and moving in with May's mother and stepfather in Palo Alto on Waverly Street. The arrangement was uncomfortable; May and her mother, Sallie, could not live together in peace, but as William still wasn't making a living they had little choice.

Bill's temper did not abate with the move. So William tried option two: the psychological method. He enrolled in a parenting course run by a Mrs A. H. Putnam at the University of Chicago, then one of America's premier schools of education, and began a long, detailed correspondence. Mrs Putnam sent several suggestions, apparently off-the-shelf strategies for dealing with difficult children. None worked on Bill, not even her suggestion they throw cold water on him, and William eventually decided the woman was of no help. They reverted to Plan C – surrender.

'Anger is about the only emotion he displays,' lamented William, 'with a little love at times.'[9]

'I've got dark eyes,' Bill told his father when he was two-and-a-half. 'I can frighten people.'[2]

They did not follow Mrs Putnam's final advice: see to it that Bill spent as much time as possible with other children.

Back in the States, and with no need to correspond with the family, William lost his enthusiasm for the diary and left large parts of Bill's childhood unrecorded. May's hoarding, especially of objects and documents related to her much-adored son, provides much of the picture of his youth.*

Their peripatetic life continued, changing homes every few months or every year, again impelled by both economic considerations and the psychological obsession for privacy. Even their family noticed. Walter Shockley and his wife, Frederica, suggested they finally settle down, for Bill's sake as well as their own, telling William and May they were turning into gypsies.[8] Gradually, they settled, remaining mostly in Palo Alto in a series of houses up and down Waverley Street in the center of town.

The Shockleys kept Bill out of school as long as they could, minimizing his contact with other children. They felt uniquely qualified to teach their son at home: May taught him mathematics and art, and his father taught him science and geography. There appears to have been little attempt at English or writing. The benefits of socializing with other children or dealing with people outside the immediate family escaped them.

When they thought they could no longer avoid formal schooling – Bill was now eight years old – they sent him for two years to the nearby Homer Avenue school where he mostly earned As and Bs (including As in deportment). 'I didn't especially enjoy school and have a vague recollection of liking Thursdays because then there was only one more day of school for the week,' he later said.[10]

William senior then sent his son to the Palo Alto Military Academy. William apparently felt the discipline would do Bill good, and judging from his report cards, which May of course saved, he was at least

---

* While a child, Shockley fell off a porch and a splinter of wood perforated his cheek, destroying a dimple. William senior measured it (2.5 inches long by 0.4 inches wide and 0.2 inches thick) and recorded the measurements; May saved it and I found the splinter in the archives.

Figure 4 Bill, in the uniform of the Palo Alto Military Academy.

partially right. Bill earned mostly As; the lowest grades were in language and spelling. He had enough of the highest marks possible in 'courtesy, neatness, promptness and good conduct' to make him an Honor Cadet, which must have amazed his parents. Bill had learned to control his temper out in the world, saving it for May and William, where it was most useful.

The school cost William $920 a year, an expense he could not easily afford. He still eked out a living on mineral investments and consulting, although May's income from selling a few paintings at $100 each helped. When, through May's connection with the Hoovers, William got a job teaching mining engineering at Stanford, the family's finances eased somewhat; still, the burden of the military academy was heavy.

May suffered one major disappointment with her son. Since 1911, a Stanford psychology professor, Lewis Terman, had been studying gifted children, hoping eventually to gather a sufficient number of subjects he could follow through their lives to see how they differed from other children. Terman hoped to prove – at least initially – that intelligence was genetic, and that the intellectually gifted did better in life.

In 1916, he began testing hundreds of children in the Palo Alto, San Francisco and Los Angeles areas using the Stanford–Binet IQ test he had recently developed. Terman accepted as subjects only those children who scored 135 or higher, his definition of genius (100 being average). Teachers initially selected who they thought were the two or three brightest children in their classes.[11]

It is not known how Bill was nominated but he was tested for the first time at the age of eight, just before he entered public school, and scored 129. The next year he was retested and scored 125. (Having a small decrease between tests was not uncommon.) He failed to make the cut. He was still two standard deviations higher than average; he just was not, according to Lewis Terman, a genius.* Later in life Bill joked often about how he could not qualify for Terman's gifted study, yet could still win a Nobel Prize in physics.[1]** That he subsequently used the same IQ tests as the basis for his unpopular beliefs about race and intelligence never seemed to vex him, nor did the fact that he was living proof the tests should not be taken too seriously. The irony was lost on him.

Terman tested May in 1919 using a different examination, and recorded her IQ at 161.

There were several beneficial influences on young Bill's life. One was his grandfather, May's stepfather, whom he loved dearly and who taught him to shoot. A considerably more pacific influence was a neighbor, Perley A. Ross, a physicist on the Stanford faculty. The Rosses became part of Bill's extended family; he was constantly in and out of their house, and Ross's two daughters became his only childhood friends. A gifted teacher, Perley Ross would explain radio theory and other problems in physics to Bill, who absorbed them attentively and then tore out of the house to play with the Ross girls.[11] Perley Ross became one of the forces that turned Bill toward physics. The greatest influence, however, remained his father. It was William, Bill thought, who had the true brains in the family. The memory of William's lessons stuck with him all his life.

'He encouraged me into scientific studies and would always discuss them with me,' Bill told a writer 40 years later. Bill remembered sitting

---

* A standard deviation is a statistically significant measurement of how much a quantity differs from the mean (average) in a statistical curve.
** Terman missed two Laureates: Shockley and Luis Alvarez, a Berkeley physicist.

fascinated as William talked about such matters as the buoyancy of gasses and the laws of Archimedes.[10] Bill later tried to tutor his own sons in the same way, but the attempts were not equally appreciated.

There was a certain self-consciousness to their relationship, mostly because of William's age. Bill remembered a 'disturbing' picture of himself less than a year old on a park lawn with his father, then aged 55, looking like a 'bearded elder,' with his trim white beard 'more appropriate for a grandfather than for the father of the baby that I was then.'[11] The affection, nonetheless, was deep and true.

Always, there was May. 'I woke up with the thought in my mind,' she wrote in her diary when Bill was eight. 'The only heritage I care to leave to Billy is the feeling of force and the joy of responsibility for setting the world right on something.'[12] Although he would deny it later, he would always gravitate back to her through his life. Even when she became a burden or an embarrassment, for Bill, there was always May.

The Shockleys lack of respect for Bill's formal education – and socialization – continued through his teens. They celebrated his graduation from elementary school in 1922 by letting Bill skip middle school. Among other things, they wanted to take a trip to London so Bill could

Figure 5 The 'bearded elder,' William Hillman Shockley.

see Victoria Street and meet his old nurse, Richmond, with whom they had kept a warm correspondence. Of the trip to London, no word remains except William's critique of London's chefs.

The original plan was eventually to move back to London, but in New York (where they were to board a transatlantic liner) William suffered a mild stroke on 8 November 1924. Fearing for her husband's health, May decided the family had to return immediately to warmer California. Then, perhaps to get away from tension with May's mother Sallie, they moved on to the Los Angeles area to a cottage at 1168 N. Edgemont Avenue.

Bill entered Hollywood High in the fall of 1924, having completely skipped middle school.

In their new home, William supervised the unpacking of his library and his collection of artifacts from around the world. He spent two months just cataloging his treasures. William also went about getting his financial affairs in order, fearing physical and mental incapacitation. He earned about $4,000 in income the previous year, mostly interest on investments, and the family was finally modestly comfortable. His estate was valued at about $75,000.

On 2 May 1925, he had trouble moving around the house, and May took him to the office of a Doctor Bowers. Six days later, riding about the city with a real estate agent, William collapsed into 'hysteria.' He had suffered another stroke. By the 18th, he felt much better, walking almost naturally. The next morning he collapsed raving on the bathroom floor. Bowers, and a specialist, examined William and told May that they doubted her husband would survive. If he did, he would be paralyzed. William gradually sank into a coma with May and a nurse tending him.

On 26 May 1925, May made a blunt entry into her daybook: '8:20. Wm. died.'[13] He was 69.

Bill was 15. Watching his father die, a man he deeply loved and greatly respected, must have been shattering. The bearded elder haunted him all his life. In 1955, in a rare moment of self-reflection, as he began forming the company that spawned Silicon Valley, he scrawled in his notebook a cryptic tribute: '30 March. Idea of setting world on fire, father proud.'

With William, the diarist, gone only May could record her family's life; and she was terrible at it. Her lack of enthusiasm for verbal – if not physical – archiving left a huge gap in the story of her son's adolescence. And produced an intriguing mystery: 'X.'

X begins appearing in May's 1925 date book, identified in no other way. All that can be said is that X was a man and his birthday was 17 June because May took him to a birthday lunch that day. Both William and Bill knew X; the most logical deduction was that he was a professional person, probably either their lawyer or a doctor, although even that is not clear. They met in X's office with William, and there was a reference to a book 'criticism' she took to him. He called often, Bill sometimes taking the message. X was there when her boy had his tonsils out on 20 June 1925. On 7 September, May made her last entry about him: 'X called about 10 minutes. 8:30.' He never appears again in any of her writing.

If X had appeared under his real name nothing about his presence in May's diaries would be interesting. But in all of her archives, in the archives of two generations of almost pathologically exact Shockleys, X is the only anonymous person. If it was an innocent flirtation or a blameless professional relationship, why the mystery? Whether May, then aged 46, was having an affair while her husband was dying, we will never know, but if she did, Bill probably knew about it.

Not much is known about Bill's high school endeavors. He was active in student affairs, but 30 years later, he wrote how lonely he had been.[15]

The week before his father died, Bill took the College Entrance Examination on 18 May 1925 for admission to the University of California. He scored in the 69th or higher percentile in the sciences, but landed in the second half in French and English courses. His score on 'Quadratics and Beyond,' was the 45th percentile. He was admitted to the University of California – Southern Branch, now UCLA. The campus then was within walking distance of his home. In 1928, he transferred to the California Institute of Technology.

The Caltech Shockley joined in the fall of 1928 was a relatively small school just beginning its march toward one of the great centers for science in the world. Located in the green, rich town of Pasadena, then still a lush exurb of Los Angeles, the school consisted of 30 acres of overgrown fields and an old orange orchard. Only about half a dozen buildings were completed, including old Throop Hall, the well-worn domed campus center. Seven years before, chemist Arthur Noyes and trustee George Ellery Hale, of what was then called the Throop College of Technology, had lured the great physicist Robert Millikan from the University of Chicago to take over the presidency of the institution, having secured a $4 million philanthropic gift. The first thing Millikan did was change the name of the place to the California Institute of Technology.

The name was something of a misnomer: Caltech was the home of pure science, engineering coming in decidedly second. From its earliest days, it drew – raided – faculty from places like Harvard and MIT by promising researchers their own labs, a free hand to run the research they wanted, and all the money they needed, an irresistible combination. Departments were encouraged to interact, a break in the hoary tradition of the more conservative eastern schools. Consequently, Millikan, Noyes and Hale attracted not only some of the brightest researchers, but the most iconoclastic, giving the school an edge of excitement, an atmosphere tinged with adrenaline.

Two years after he arrived, Millikan won the Noble Prize for his study of charges and the photoelectric effect. His eminence further drew illustrious scientists from all over the world to visit, lecture and stay. Edwin Hubble began his masterful analysis of the galaxies at Caltech. Albert Michelson, having won his Nobel for measuring the speed of light in Cleveland, repeated the experiment in Pasadena. Linus Pauling was a young faculty member who would eventually win two prizes, one for chemistry and the Peace Prize, the only person ever to win two unshared Nobels. Einstein came to visit and lecture twice.

Millikan hired Charles Tolman, chemist, physicist, outdoorsman, philosopher. He brought in Hendrik A. Lorentz from Germany and Charles G. Darwin, the grandson of the biologist and a respected mathematical physicist from England.[17] The faculty club, the marvelous Atheneum, opened during Shockley's second year and became the hostel for the world's greatest scientific minds. Einstein stayed there, eating and kibitzing in the plush dining room. Physics was entering its golden age and if you wanted to see the movers and shakers you would do well just to hang out in the lobby of the Atheneum.

Most of the students commuted from homes in the Los Angeles area; the few graduate students who were not local took up the scarce dorm rooms but mainly rented rooms in the neighborhood. There were no undergraduate dorms. At noon the school emptied as the commuters, by car and electric trolley, went home.[1]

Shockley loved the class structure, created by Pauling, which differed greatly from his experience at UCLA. The school grouped students into sections, 15 or 20 to a section, who attended all their classes together. All freshmen (despite his year at UCLA, Shockley was considered still a freshman) and sophomores took more or less the same courses no matter what their majors. Only in the junior year did students begin to specialize. Shockley felt it gave every Caltech student a firm basis for the courses that followed. The exception was physics. There, he found himself in one of the two honors sections that provided accelerated instruction. The competition to get into those sections was fierce and the competition within the sections even worse.

By this time, 1928, Shockley had grown to his full height of 5' 8" and weighed about 130 pounds with a sleek body of muscle and bone that he nurtured with constant exercise. He was fit and handsome enough to pick up extra money modeling for a sales pamphlet published by the manufacturer of the Trelor Strength program. Shockley is shown in a series of photographs doing calisthenics with Trelor tension devices. He would never be far from a gymnasium or a swimming pool the rest of his life, and never weighed more than 150 pounds.

His blue eyes had darkened even more, turning flinty steel gray. His brown hair was just showing the first signs of thinning. He had turned into a quick-moving bright young man, sure of himself, eager to get going, and aggressively competitive. He carried that sense of

Figure 6 Shockley advertising Trelor Strength equipment during the Caltech years.

competition with him at Caltech. It became fuel for both his meteoric rise and his disastrous fall.

He had by now also gained two other aspects to his personality that he carried happily until the world around him got too serious. They ensured that no matter where Bill Shockley worked or studied he would be impossible to ignore.

Sometime during high school he learned magic tricks and became an accomplished amateur magician, very popular at parties, usually the first adult invited to children's birthdays. This ability also made him dangerous on a speaking platform in later years: he would introduce a speaker, reach over to shake his hand and a bouquet of flowers would blossom from his sleeve; or worse, a live dove would erupt from his jacket and circle the room in reconnaissance. He was picked as discussion leader at meetings or as master of ceremonies at some risk to the hosts.

To this he added an extraordinary talent for practical jokes. His colleagues learned to expect the worst. Nonetheless, some of Shockley's greatest stunts are still spoken of around the campfires at places where he studied and worked.

Caltech was exactly the right place for Bill to exercise his talent for magic and mischief. He earned spare income doing magic at parties, and practical jokes have always been considered an art form there. Dismantling huge restaurant signs overnight and transplanting them to the campus is considered child's play by students; exploding devices at the mid-field line during football games became a cliché in later years, even when the game did not involve Caltech's talent-challenged, near-sighted football team.

Shockley undoubtedly participated in some of the traditional stunts, but his endeavors tended toward the more subtle and complex, relying less on explosions and flashing lights. Fifty years after the event, for instance, grads still remember Shockley's finest performance: the adventures of 'Helvar Scavi.'

Fritz Zwicky was a Caltech physics professor of considerable ability. He frequently ended a class by putting a complicated formula on the blackboard and told the students: 'This is the answer. Now come back and tell me the question.' German–Swiss, with a heavy German accent, Zwicky was capable of being unkind. Before dissertation defenses, students often visited professors on their examination committee. The professors would tell the students in general what questions they expected to ask during the oral exams. Zwicky sometimes lied, trapping the student with questions that in no way matched what he actually asked. Graduate students were not amused. But he had one weakness: he never had any idea who was supposed to be in his class.

All this made him a perfect mark for Shockley. Using a blank registration card, Shockley enrolled a completely fictitious student named Helvar Scavi in Zwicky's class one autumn. At the mid-term exam, traditionally an open book test, Shockley arranged for an extra student to show up in the class. With no tally of registered students, neither Zwicky nor his teaching assistants noticed. When the test was distributed, the student took a copy and slipped it out the window, where another student ran with it to a nearby office. There, Shockley had gathered a group of students who had already taken the course, as well as a number of faculty members who, of course, knew the answers. According to legend, one was an unusually gifted young physicist named J. Robert Oppenheimer. They answered the first few of the five questions perfectly and then scribbled in: 'Well, I'm too damn drunk to write any more.'

The test book was slipped back to Zwicky's classroom and turned in.

This continued through the year, with Scavi answering most of the questions brilliantly, not filling in a few just to confuse the professor. Zwicky, who never caught on, was amazed at Scavi's brilliance.

Zwicky got revenge – unintentionally. Marking on his version of a curve, he gave Scavi an A and every other student in class a C–.[18]

The humor acted as a release from the pressures of what had become one of the most serious science institutions in the world. To keep competitive, Shockley found himself going back to Stanford every summer for additional courses, particularly in physics and especially when Perley Ross was teaching.[19]

Two professors at Caltech appeared to have had the most influence on Shockley. William Vermillion Houston, later president of Rice University, gave the introduction to theoretical physics. Houston was a superb teacher whom Shockley remembered all his life. The other was the imaginative Richard Tolman. Tolman's vision extended as far as science could see and comprehend. It was his work – and Millikan's money – that brought Einstein to Caltech in 1931.

Houston, Millikan and Tolman involved Caltech in one of the greatest revolutions in the history of science, a profound intellectual battle that transmuted physics. These men – especially Millikan – made Caltech an active center for the coming encounter. Shockley had a clear view of the action.

The battle raged over nothing less than humanity's view of nature. Every physics textbook was being rewritten, seemingly every other month. Millikan and Tolman were deeply involved, having bet their reputations and the reputation of Caltech on the young rebels then joyfully revolutionizing their science. Nothing could be more profound or seductive than the intellectual tumult that engulfed the campus. At Caltech, Copenhagen, Heidelberg and Göttingen, physics and existentialism became one.

Shockley was swept up in the maelstrom.

CHAPTER 2
The lightness of being

The revolution of physics that Bill Shockley watched at Caltech not only undermined Newton's orderly, rational and predictable universe, but stemmed from one of the most intriguing and beautiful arguments in physics: What is light? Is it a wave of something or a stream of particles?

Newton was the first to declare for particles, in 1704. He was contradicted 95 years later by fellow Englishman Thomas Young, who demonstrated that if you superimposed one beam of light on another you could produce bands of light and darkness. The only logical explanation was that light had to be a wave – if one wave interfered or overlapped with another, they blocked each other out. According to James Clerk Maxwell in the mid-19th century, light waves and electromagnetic waves are simply different facets of the same thing, both traveling at the same speed. One hundred years of experiments resolved the issue in Maxwell's favor: Light consisted of waves.

In the first year of the 20th century, a German, Max Planck, added heat to the discussion. Maxwell described heat as what happens when moving particles in a gas collide and give off energy: The faster the motion of the particles, the more collisions, the greater the energy released, and the more heat produced. He had equations to describe all of this. Experiments showed that the equations worked well at higher temperatures, but failed to describe what happened at the coolest. When laboratory physicists find that their experiments do not follow what the theorists say should happen, there are two possibilities: either they are doing the wrong experiments (or doing them in the wrong way), or the theory is wrong. Here, no one questioned the experimenters.

In a furious few weeks, Planck produced new equations to explain the experiments. He resorted to a mathematical fiction: He explained the contradictions by assuming that the gradations in radiation did not go in a smooth, gliding path, but jumped in discrete steps, which he called quanta. The number that determined each quantum leap in his equations is constant.* He called it h, forever after known as 'Planck's Constant.'

The theory did not seem to make any sense; did nature really work by jerks?

Planck strongly disliked his own theory and was almost apologetic in presenting it to a physics meeting in 1900, and in a subsequent paper. His colleagues were unimpressed – except the 26-year-old patent clerk Albert Einstein, then living in Switzerland. In 1905, Einstein published a paper in *Annalen der Physik* – his first major paper – showing that Planck's theory was not that weird.

If light and matter operated within the same paradigm, how could Maxwell find a smooth flow (probably waves) and Planck find jerks (which indicated particles)?

Einstein believed that part of the discomfort Planck and his fellows felt was that physicists thought of matter as particles – atoms – while light and electromagnetism were thought of as waves. If physicists perceived light as particles, or more precisely, packets of particles, this conflict is resolved. The smooth flow that Maxwell mistook for waves simply was a blurring caused by the particles zipping by too fast to be seen clearly.

This was the proverbial stroke of genius, and if Einstein had done nothing else in his life his fame would have been assured. He went further: He showed that these quanta explained several known phenomena, most particularly the photoelectric effect – that when you shine light at a substance it emits electrons. If light really was a stream of quanta, increasing the quanta bombardment frequency would naturally blast off more electrons. It was for this, not for his far more radical and controversial theories of relativity, that Einstein won his Nobel Prize.

As is common with new theories, this one had holes in it and Planck and others had little difficulty finding them. Einstein filled most of the gaps himself by suggesting that quanta existed not only within light but also within all matter as inner vibrations or heat. Einstein's problem was that the laboratory physicists then had no way to prove or disprove his theory, which kept his colleagues from accepting it.

---

* A quantum leap is actually very small. Language has distorted the meaning.

At this point, around 1912, Millikan, still at Chicago, came to the rescue. After ten years of trying, Millikan found experimental proof of Einstein's theory. That work later won Millikan his Nobel in 1928, the year Shockley entered Caltech.

Light quanta became known as photons in 1926, so named by a British physicist, Gilbert Lewis, and the idea that photons existed slowly worked its way into the consciousness of the physics community. Then things moved into the existential.

In 1911, Ernest Rutherford, working in Cambridge in the UK, had described the atom as a miniature solar system: a number of electrons orbit about a heavy nucleus just as planets orbit the Sun. Most physicists concurred because Rutherford based his work on commonly accepted experiments. Unfortunately, this view violated Maxwell's equations of electromagnetism. Maxwell's vision would make such an atom too unstable; the electrons would plunge into the nucleus rather than remain in orbit.

Scientists also were discovering other things in the atom that contradicted Maxwell: sharp bands showed up in their spectroscopes, also impossible in Maxwell's atom because that seemed to indicate incremental changes in energy as opposed to a smooth flow. That paradox haunted them.

Enter the Danish physicist, athlete and philosopher Niels Bohr. In 1913, Bohr published a paper saying that the electron orbits in a hydrogen atom (the simplest one) were restricted to prescribed paths at set distances apart. Instead of sailing around the nuclei and shifting orbits smoothly, the electrons jerked from one orbit to another when you added or subtracted a packet of energy. If they lost energy, sometimes as light, they fell back to a lower level orbit. Add energy and they jump up one. Only when an electron jumps from one orbit to another, said Bohr, can it give off a photon. He showed how this fit with Planck's theory, even using Planck's h to describe the ratio of energy change to the frequency of light. The theory also explained why burning elements gave off light of precise wavelengths, a fingerprint of their electron orbits. Bohr won his Nobel Prize in 1922.

So atomic contents took Kierkegaardian jumps with light as a physical manifestation of those existential leaps. Since the dour Danish philosopher heavily influenced Bohr, Bohr had no trouble making such a leap himself. The German physicist Arnold Sommerfeld, influenced by Einstein, added to the picture whirling ellipses of electrons like planets sailing through the solar system. It became known as the Bohr–Sommerfeld model.

Tolman used Sommerfeld's new book on atomic structure in the class Shockley took at Caltech, going through it page by page with his students (he hadn't seen it before either), trying to unpeel the atom with them. The rapid pace of discovery upset the normal teaching hierarchy: teacher and student learned together.

But the Bohr–Sommerfeld model had its own problems. How could electrons jump from one orbit to another without existing for however brief a moment of time somewhere between those orbits? Further, spectral readings for the most complicated atoms did not support the model.

Discoveries tumbled upon discoveries. A French nobleman, Louis Victor duc de Broglie (Nobel Prize 1929), came up with a theory that tried to bridge the gap between the particle exponents and the wave advocates. What if light really consisted of a stream of particles accompanied by a wave? Different experiments produced different results because they measured two sides of the same thing. Einstein was one of the few scientists who thought this notion plausible.

The Austrian Erwin Schrödinger took Einstein's word for it and in 1926 developed a new theory called wave mechanics for which he won the Nobel Prize, shared with Paul Dirac, discoverer of anti-matter, in 1933. Schrödinger's atoms didn't fly around the nuclei like little planets circling a star; they were fuzzy clouds that swaddled the nuclei, vibrating at a certain frequency. The electron isn't in one place; it is simultaneously everywhere in its orbit around the center of the atom. Only certain stable frequencies are possible, these describable only by a whole number. Add energy and the electron jumps to the next whole number frequency. Remove energy and the reverse happens. Between those two numbers, it doesn't really exist.

A year later, the year before Shockley entered Caltech, physics went completely mad. Fomenter of the chaos was the 25-year-old German Werner Heisenberg working in Leipzig and with Bohr in Copenhagen. Heisenberg attacked the very basis of his own science, adding a level of existential doubt that Bohr found compelling and Einstein repulsive. It bagged him a Nobel in 1937.

Heisenberg said that the physical model of the atom, Bohr's as well as Schrödinger's, was impossible. You had to think of an atom as a mathematical construct, not as a physical entity. Heisenberg said that we can know an electron's position, but if we do, we cannot possibly know anything about its motion. Or we can describe its velocity, but if we do, we give up the possibility of knowing anything about its location. The electron existed in reality only as a mathematical probability – it was probably here, it was probably moving at this speed. Even studying an atom alters it: You can't look at an electron without using light and the light knocks any electron out of position. Moreover, predictions are impossible. The best we can hope for, Heisenberg said, was to describe probabilities: the atom is probably moving with certain momentum or velocity, or the atom is probably here, not there. We can't even demonstrate causality, since we cannot prove where the electron was or how fast it was moving.

Imagine the excitement of being a student in physics while all this was going on. Wednesday's truths were replaced by Thursday's revelation. Imagine the ramifications of the theory, known to laymen as Heisenberg's Uncertainty Principle and to physicists as the Principle of Indeterminacy. Heisenberg framed reality as nothing more than random events and probabilities.

Things got even worse, when Bohr, later in 1927, took another step in the direction of chaos with his Theory of Complementarity. Bohr said it was impossible even to describe what happens in an atom in the common language. The atom lives in a quantum world: we dwell in the mundane non-quantum or classical world, where a dropped ball falls and the teapot stays where you left it. A physicist doing an experiment does so in the non-quantum laboratory using non-quantum equipment and a non-quantum state of mind. He studies quantum nature, but when the experiment is over, he must explain what he saw in non-quantum terms. The only way to do that is to assume that definitions in the vernacular fail.

Is light a wave or a stream of particles? Bohr answered: 'Whatever.'

Einstein would have none of this. Light could not be both wave and a particle. It had to be one or the other. The material world could not be simply probability; an electron had to exist in reality, it had to be somewhere, moving at some speed and we ought to be able to spot it and measure its motion. Reality was reality. He attacked with a stream of arguments, but he was trapped in the language. Bohr and Heisenberg had defined the vocabulary and Einstein's language could not supply relevant definitions. They were speaking different tongues, inhabiting a different universe. He proposed a long line of ingenious experiments to prove that there was a reality beyond the experimenter and all were rebuffed. Schrödinger pitched in on his side with his famed and hilarious thought-exercise, Schrödinger's Cat, which drove quantum proponents to distraction, but he too eventually failed. He and Einstein were the last two major holdouts.

Was this physics? Einstein did not think so. Metaphysics was more like it, perhaps solipsism.

What they fought over, of course, was philosophy as much as science. Einstein lived in a universe of whirling things and powerful forces, where everything affected everything else and we could watch it and draw conclusions because the universe was rational and so were we. Although Einstein had overthrown Newton's construct of nature, Einstein accepted his deterministic view of the process of nature. Mathematics describes reality; it is not itself reality.

Millikan thought Einstein was wrong and so did Tolman. Shockley would sit in Tolman's office and talk about the battle. He would watch Einstein's ambling through the palm groves from the Atheneum, and listen to reports of the conversations. Millikan brought Einstein to Caltech partly to convert him to the new paradigm. Bohr traveled later in the decade to Einstein's new home at the Institute for Advanced Study in Princeton, hoping also to convince the great man that he was finally wrong. He didn't.

Before the beginning of the 20th century, decades had elapsed between major discoveries. Now they were separated by years, sometimes months, occasionally weeks or days. The telephone would ring – often at Tolman's office or Millikan's – reporting that another amazing thing had happened. Physics students such as Shockley were swept up in the hullabaloo. It became the main topic of conversation and affected their studies (forget the textbook, it's wrong), their view of their discipline and, for many, the work they would embrace in the future.

Shockley's world changed in the four years he was at Caltech. The certainties of Newton and Maxwell gave way to an indeterminable nature ultimately less physical than it was mathematical, where science was limited by what it could ever know. Quantum physics laid waste the whole idea of the atom, with its electrons and nuclei, and begot quantum mechanics, a merger of Heisenberg, Max Born, de Broglie and Schrödinger.

Shockley began to probe deeply into quantum mechanics. His directed reading project was a book on the subject by the British physicist Paul Dirac. He took a summer course on Bohr orbits at Stanford

Figure 7 Shockley, 1930.

taught by Karl K. Darrow, nephew of the famous lawyer and a physicist. Darrow's cool, precise, colorful lectures drew Shockley into the roiling currents of the new physics.[20]

What good any of this would do, what effect it would have on the ordinary, non-quantum world was not something Shockley could then imagine. Tolman, Millikan and Einstein didn't know – how could they, embroiled as they were in monumental intellectual debates over just how random the universe was?

The fact remained that one could not understand how electricity (the flow of electrons) is conducted without understanding this quantum universe. Shockley absorbed it all when he launched off in his non-quantum DeSoto convertible to follow his ancestors into the Massachusetts Institute of Technology in the fall of 1932.

—≺—

Shockley had applied to MIT and Princeton, but heard from MIT first and accepted days before Princeton offered him a place. His faculty advisor in Cambridge was the physicist John C. Slater. Slater started their relationship off on the wrong foot, sending Shockley a warning letter. Slater had heard that Shockley would do very well in subjects he was interested in and blow off those he didn't care for. That would not do, Slater cautioned. Shockley admitted the accuracy of Slater's appraisal and thought it must have originally come from somebody who knew him well. He suspected the source was a Caltech acquaintance, Frederick Seitz.[20]

Seitz, a San Franciscan a year younger than Shockley, was the son of an immigrant baker from Germany and a woman of colonial American stock whose father, a Union Army veteran, went west after the Civil War. Seitz attended one of San Francisco's best private high schools, Lick-Wilmerding, founded by the eccentric philanthropist James Lick, who made his fortune in South America and in the California Gold Rush.

Seitz was admitted to Stanford University as an undergraduate, and majored in physics under the tutelage of William Hansen and Perley Ross. '[Stanford] did a good deal of research and Hansen gave a course on quantum mechanics which I took as a junior. It was clear and explicit enough so that one had little difficulty in following it.... These people were all familiar with quantum theory, with the Bohr theory of the atom. People talked about the wave equation as though it were a believable thing in physics. So the mood and atmosphere, though somewhat erratic, was pretty solid.'[21]

At Stanford Seitz also met the 26-year-old Edward Condon, a star of the ascending physics department at Princeton, then a visiting lecturer. The two became friends and Condon suggested Seitz apply to Princeton for his graduate work.

In 1930, Seitz transferred to Caltech for a year to pick up depth that was more technical.

Seitz, too, flourished in the academic atmosphere created by Pauling at Caltech, and in a theoretical physics class taught by William Houston he met Bill Shockley. By that time, Seitz had heard a great deal about him. He had spent considerable time with the Rosses at Stanford. Shockley's frequent visits to the Ross house and his friendship with the Ross girls made him a constant presence there. May also visited the Rosses often. The Stanford physics department was quite small, Seitz remembered, much like an extended family, so he couldn't avoid hearing of the Shockleys.

With the influence of Condon, he was accepted at Princeton.[1]

Seitz returned to San Francisco in June of 1932 and spent some time with friends at Stanford, including the Rosses. One day, Mrs Ross mentioned to Seitz that Shockley had been admitted to MIT for graduate school and was going to drive his DeSoto to the East Coast. He was looking for someone to share expenses. Seitz called Shockley and in August went down to Hollywood to spend the night with the Shockleys so they could leave from there.

Off they sailed with the top down through the shimmering summer heat. Seitz's only concern when they got into the car was that Shockley had brought one of his grandfather's pistols with him, loaded and in the glove compartment.

Shockley soon stopped shaving, and with his dark hair covered with a beret, must have looked quite alien on the highways of the sad, dirty western and Midwestern countryside, then still caught in the dark embrace of the Great Depression. They took the southern route, through Arizona, New Mexico, Texas and Arkansas, hitting the Lincoln Highway (Route 40) in Ohio. The weather ranged from baking to steaming to pouring.

Traveling across the Kentucky hills one early evening, the two men nearly ended their illustrious careers prematurely. Seitz was driving on a narrow two-lane road with a sheer drop-off on the right. 'As we rounded a curve, we saw hurtling down toward us two trucks, clearly racing one another, taking up both lanes,' he wrote. 'Apparently chasing each other up and down the mountains was a popular sport among the spirited, young locals.' Seitz darted to the right, hoped there was just enough room on the shoulder to avoid the oncoming trucks and precipice, and was safe by mere inches. The trucks roared by without so much as a horn honk. 'To the best of my knowledge I have never been closer to instant death than in those few seconds.'[21]

Shockley was also taken by speed, and that would not be the last time he was nearly killed on a highway. Fortunately for both him and Seitz, the old DeSoto could only do about 50 miles an hour, and then downhill with a little help from a tail wind.

By the time they reached New Jersey, Shockley apparently looked suspicious enough, with California license plates on the car, an unshaven face and a beret, to attract the attention of the Jersey City police, who stopped the car and found the loaded pistol in the glove compartment. He wired May not to worry if she heard he was in jail.

Fortunately for all, that didn't happen. To get the gun back he had to get a letter from the police chief in Cambridge, arranged through MIT (he also immediately got a gun permit on arrival in Cambridge).

The two young men chatted both about physics and life in general on the trip east. Seitz was taken by several characteristics of his new young friend. Seitz quickly discovered that Shockley was 'unusually intelligent.' He could look at a scientific problem and immediately bore in on the core. In his experience, the only scientist he met better at this than Shockley may have been Enrico Fermi.

He also noted, in retrospect, a kind of elitism about Shockley's philosophy. 'He was inclined to believe that society should be governed by a vaguely defined intellectually elite group, rather than by majority rule as in a democratic society,' Seitz wrote a half a century later, and after Shockley had achieved considerable notoriety. 'I saw an inkling of this... but I did not take it seriously then.'[21]

Shockley and Seitz finally pulled into Princeton on a moon-drenched night and Shockley dropped Seitz off at Condon's barn-like house and continued north to Massachusetts. The two men corresponded and visited each other through the year.

They remained close friends for almost as long as they would be blood enemies.

Shockley found a room at 6 Cleveland Street, about 20 minutes walk from MIT, with the help of fellow physics student John Potter. The room cost $4 a week. He and Potter talked the landlady down from $4.50 on the ground that he had to pay $3 a month to park his car in the backyard across the street. That brought his living expenses to $20 a month, including rent and parking, a little less than the $250 to $300 a year that the dorms charged. His stipend from MIT was $80 a month.

The room on the second floor to the left of the front door was large, about nine feet by twelve, with cross-windows. The neighborhood was Irish, as was the landlady. The house was two stories, wooden, painted dark with white trim. Potter took a room on the first floor.[22] 'The room is not bad at all,' Shockley wrote May. 'I do most of my studying in my office anyway. Robert Richtmeyer and Smythe from Belfast, Ireland (Queen's University) and a chap named King share the office with me.'

Because money was tight – May had not yet recovered from the stock market crash of 1929 – Bill took an unusually heavy teaching load that cut in on his research time. He had three sophomore laboratory sessions, which took six hours out of his week, one lecture (he called it a 'recitation'), an hour with about 16 lab papers to grade, and his own class work.[23] His workload increased in the second semester. He took four courses and taught 12 hours. The courses added up to 39 course units, and the teaching, with about two hours preparation, added another 14. Even by MIT standards, that was a lot of work, but he could not afford to dawdle. Because of the Depression, even MIT was having problems. All the graduate students took a stipend cut, in Shockley's case, down to $77 from $80, 'because the Institute is getting too far in the red.' He promised his mother to cut expenses even further at the end of the year.[24]

Shockley had a serious problem, however. He and Slater did not get along. Slater's intellect wasn't the question and he was reputedly a very good teacher, 'very precise.' He was not, however, a good graduate advisor. He was distant and unhelpful, Shockley said, and he learned how not to deal with graduate students from Slater, which included doing too much for them.[20]

Ironically, years later the two men became good friends and colleagues, and Shockley eventually found a substitute mentor, one of the best things that ever happened to him.

Shockley also found a friend and kindred spirit in his first year at MIT, James B. Fisk.[25] Like Shockley, Fisk had a squarely middle-class background, coming from a long line of lawyers and judges. He went through public schools in Rhode Island, including a technical high school that prepared him for such things as carpentry and machine shop. It also prepared him, apparently, for MIT. Fisk became Shockley's alter ego and ally in mischief.

At the end of the spring term, Seitz's friend at Stanford, Bill Hansen, who had received a fellowship at MIT, joined Shockley and Seitz for the drive west. They detoured to Ann Arbor to visit the University of Michigan, then a great center for theoretical physics. Hansen stayed on at Michigan, and Shockley and Seitz continued west. They parted in Salt Lake City, Seitz continuing home to San Francisco and Shockley to Los Angeles.

In August, Seitz called Shockley to see about the return trip, but Shockley told him he would have to make plans on his own. He was driving east – with his wife.

—<

There is no reference in any of the voluminous Shockley family correspondence to Jean Alberta Bailey before Shockley married her in August 1933. Seitz said Shockley never mentioned her in conversations: It appears Shockley told no one he had a girl friend or was engaged or was thinking of getting married. None of his remaining letters to his mother mentioned her. No one knows how and where they met or why they got married.

Jean was born on 13 June 1908 in Cedar Rapids, Iowa – making her a year older than Shockley – the daughter of Albert and Anna Condit Bailey. Bert, her father, had hoped to be a Christian missionary but failed his physical examination when the doctors detected a heart murmur. He took a job instead, teaching ornithology at Coe College in Cedar Rapids. Anna was college-educated – she knew Greek grammar – sickly, shy and retiring. Bill remembered her as a woman who spent most of her time reclined on a couch. Jean learned about religion and birds from her parents, interests that lasted her lifetime.

Jean had two sisters, one six years older, the other six years younger. When Bert died – Jean was nine – Anna moved her family to Los Angeles to be near her brother, Ralph Schroeder, an artist. Jean met May

Figure 8 Jean Alberta Bailey in California about the time of her marriage.

Shockley through Schroeder, and presumably Bill Shockley. She was studying at UCLA at the time.

Small, slim, about five-three, Jean was a pleasant-looking woman with blue eyes and a round face with high cheekbones, which made her look slightly exotic when she was young. She wore her brown hair short when it was stylish in the 20s, but let it grow long when the fashion allowed, eventually braiding it in the back. The only physical disappointment, she said, was her hands, squarish and stubby, not graceful as she wished.[14]

Jean was educated and intelligent, perhaps better read than Shockley, who had little time for anything other than physics. 'I think she was a pretty sweet person,' her daughter said later, 'though hardly glamorous in a Hollywood sense.'

'Everyone loved Jean,' said Fred Seitz.[1]

Not everyone.

Whether Shockley and Jean were in love then is unknowable, but they certainly were in lust. After one particularly passionate evening in the back seat of May's Buick, they spent two 'naked evenings' at his home, which included a romp in the bathtub, and a day or so later, he wrote Jean a letter on MIT stationery he apparently never mailed. He said he felt like an old man who 'had done everything he wanted and sat and chuckled over his memories and at other people.' Jean had been a virgin. It is not clear from the letter what Shockley's previous experience had been.

The Buick wasn't bad but it did prohibit really proper relaxation afterwards. 'The idea of resting on your breast... while the stiffness goes out of my penis and slips out of you just by itself appeals tremendously. Then it would be nice to roll off just enough so that we could both sleep comfortably. That's another trouble with cars; you have to go places and undress before sleeping, instead of just dozing off.'*

---

* I found the letter in an envelope marked 'To Be Destroyed Unopened,' one of two such in a safe in Shockley's home. Jean's name is not on it, nor is there a date. However, it was written on MIT Physics Department stationery, which places it after the fall of 1932. References to 'home' make it likely he was referring to Los Angles, not his rooming home in Cambridge (an unlikely site for a nude bath with a young woman). May owned a Buick. Following their marriage at the end of the summer of 1933, they spent almost no time back in California, so the months before the marriage would be the only time this incident could have happened and Jean was by far the most likely known inspiration for his erotic musings.

He went on in lusty detail about his pleasure and his pleasure in giving her pleasure. 'I wish I had all of you here with me to do with as suits us both best,' he ended. 'Good night at a distance, dammit.'

He was 23, she 24.

Jean was probably pregnant when they married and it may have been the reason for what obviously was a quick decision. They honeymooned on Catalina Island and afterward packed what they could into the DeSoto and headed on a direct line east, sending telegrams back to May every day, letting her know how they were doing. May began shipping presents east the moment the two left California; the Shockleys arrived to find a pile of gifts waiting for them. They found an apartment when they arrived back in Cambridge in September 1933, two rooms, a kitchenette and bath, all furnished. They paid $46.50 per month on a nine-month lease, plus gas and light; water, heat, electricity for the refrigerator were 'on the house.' A photograph of their apartment shows a wooden floor, bare, well-designed wooden chairs, a small folding table in the kitchen and two candelabra. When they moved in, they found that kitchenware was not included in the word 'furnished,' as it was customarily back in Los Angeles, so Shockley had to appeal to May to send household supplies.[26]

The Shockleys lived in almost half a dozen apartments while Shockley was in graduate school. They got by on $130 a month, which included income from his MIT stipend and a part-time job Jean managed in a nearby department store, plus continuous small checks from May. They parked the DeSoto on campus to avoid high parking fees. Jean settled into domesticity with great ease. Housework wasn't as much fun as she thought it would be, but 'I love to do it, and I really don't consider it boring in the least.' She liked the physical activity, even with a broom in hand, and 'I am firmly of the opinion that somehow, one's capacity to get things done expands in proportion to what you have to do....'[27]

Housework was occasionally interrupted by Shockley's giving his new wife a crash course in physics, which she said she loved. He was studying the electron configuration of atoms for his doctoral exam and found that he could get some order out of what he was studying by discussing it with Jean even if she really didn't understand much of it. He was prepared to teach her what she needed to know. 'I suppose I shall always be exiled from the mathematical aspects of his work, but some of the physical concepts weren't incomprehensible without years of math and physics, thank goodness,' she wrote in her constant correspondence with May.

Shockley and Jean were still keeping her pregnancy secret as Christmas neared. Shockley suggested several presents May might send her without mentioning her greatest need: maternity clothes and equipment for the baby. A month later, they apparently decided they could keep the secret no longer and told her.[28] May was not pleased. Not one to worry about things out of her control, she quickly recovered and energetically discharged another barrage of packages to the Shockleys: maternity clothing, housewares and California delicacies. The Shockley household soon became the generous distributor of artichokes to the MIT community, which made them extremely popular in the physics department. Jean worried that May was depriving herself.[29] The plea fell on unlistening ears: the packages continued unabated.

The couple fitted in immediately with the social life at MIT. Jean joined the Tech Matron's tea circle and 'poured' when the physics department's turn came to host. They had large dinner parties – or as large as their apartment would allow – with guests from Shockley's department. Phil Morse (Fisk's advisor) and his wife were favorites, with Shockley and Morse spending hours on the couch talking physics. Shockley celebrated his birthday by writing physics equations all over his birthday cake with an icing bag stuck in his mouth. They did huge jigsaw puzzles with company, or together when the company left. By all appearances, Shockley was a happy man.

'In spite of his hard work,' Jean wrote May, 'I don't believe Billy is feeling run-down. He does need to catch up on a little sleep, but the dimple seems to be in good working order, and as long as it pops up at frequent intervals, I feel content.' Shockley had only one dimple, on his left side; the place the matching one should have been on the right was marred by the scar from his youthful fall off the Palo Alto porch.[30]

The pregnancy went easily. Shockley regularly stayed up late working, but managed to sleep in on weekends and during school breaks, apparently detesting having to rise early for school. The Shockleys lounged in bed on the weekends, she in her negligée, and he in his shorts. She read, while he lolled under a sunlamp, a practice he continued through most of his life away from California.[31] 'So far I haven't felt that Billy has felt the strain of a prospective baby,' Jean wrote to May. 'I suppose that will come later with the baby....'[32]

Jean went into labor at the end of March, several weeks early. She couldn't wake her soundly sleeping husband, so she phoned the doctor who told her to go into the hospital. Shockley resisted the news and seemed reluctant to get up, so she announced loudly where she was going and inquired whether he wanted to come along too. Eventually, he rolled out of bed. They took a taxi to Baker Memorial Hospital in Cambridge. As was the norm in Boston-area hospitals at the time, they immediately gave her a general anesthetic without asking and she woke up early Sunday morning to the news they had a daughter.

The delivery was normal – no forceps – and the baby was doing well, although, as was also customary then with premature babies, Jean did not see her for several hours. Only after making a fuss in the morning did she get to introduce the little girl to her father, if only for two minutes. 'She is just precious,' she told May. 'Of course she didn't open her eyes for us, but she has the dearest shaped little head, the sweetest mouth.' The baby had blue eyes, they soon discovered, and dark hair. They named her Alison after the subject of what Jean called the oldest known English love lyric.

Shockley brought a bouquet of talisman roses to the hospital; friends sent potted plants. Shockley distributed artichokes instead of cigars, to the continued joy of his colleagues.

Jean wrote to May, 'I'm really thrilled at the prospect of being a good cow, however, which seems to be probable today. I'm awfully juicy and delighted about it... One soon gets quite unblushing about things, which is fortunate, for a maternity ward would be a difficult place for a lady with too delicate a sense of personal modesty!'[33] How May, who numbered childbirth as one of the most disgusting events in her life, reacted to her juicy 'good cow' daughter-in-law is not recorded.

They settled into family life and Shockley flew into his exams. 'He seems as frisky as ever,' Jean reported, 'and I don't believe the ordeal, if it can be called that, is bothering him at all.' He ran on adrenaline before the test, even singing in the bathtub in the evening, and when it was over, suffered from a depression when the adrenaline shut down.[34]

'Now I never have to learn anything any more. How dull!' he told Jean. Of course, he still had a small matter of a dissertation left.[35]

'The baby? Oh yes, we have a baby!' Jean wrote May in the middle of it all. 'She is not just a baby either; she is the most remarkable and lovely phenomenon in the world – our world, that is.' Jean extolled everything, from the baby's consumption of cod liver oil to hiccups, along with paeans to long naps outside in the warm spring afternoons.[36]

In July, Shockley had a scare. The night of the seventh, he began having gas pains. He took some soda to calm his stomach but he was restless and sleepless all night, with a mild temperature. The infirmary doctor the Shockleys used was on vacation and the symptoms didn't seem particularly serious, but some instinct warned them this was more than a passing episode and the substitute doctor suggested Shockley get to Brooks hospital as quickly as possible.[37] Jean left Alison with a neighbor and joined him. Shockley's appendix was about to burst, and the cautious physician and quick surgery saved his life.[38] He recovered without mishap.

—≺

Sometime during his life at MIT William Shockley acquired a curious affectation: He decided his signature from them on would be $W = Shockley$ or $W = S$, and that is how he signed his name for the rest of his life. No one knows why.

He published a scientific paper with Fisk. 'We did a little work together,' Fisk said, 'we knew each other and we both had some odd characteristics, like we were always interested in practical jokes.'[25]

One morning, Karl Compton, the MIT president and a distinguished gentleman, got into the new, automatic elevator in his building to take him up to his office. He pressed a button and the elevator ascended to the wrong floor. Assuming he had pressed the wrong button, he tried again. Again, the elevator went to a wrong floor. He saw a student and asked the student to push the correct button and the car went to yet another floor, wrong for the third time. Shaking his head, Compton got out and walked.

The previous night, Shockley and Fisk had broken into the building, unscrewed the panel covering the electric controls and rewired the elevator. Whether Compton was the target or just a lucky catch will never be known.

One result of Shockley's teaching was that for almost a decade a substantial number of the experiments used in the freshman physics lab were his design.[20]

The MIT graduate students worked out an informal exchange with their colleagues at Princeton and the two groups visited each other by train or they drove when they had the time. At Princeton, they listened to lectures by Eugene Wigner and met Einstein. They drew close to several graduate students, including Seitz.[25]

At Slater's suggestion, Shockley concentrated his dissertation efforts on studying the actions of waves of electrons through sodium chloride, a

study that produced at least three scientific papers, unusual for a doctoral dissertation. Seitz, working with Wigner, had developed the technique for calculating wave functions in a crystal containing metallic
sodium, and Shockley applied their method to sodium chloride – a com-
pound. The result was the first realistic picture of energy bands in a com-
pound crystal.[20] Mostly what he said he got out of it, however, was the
discipline of sitting at a desk and doing the laborious hand calculations.

By this time, Slater and Shockley had agreed to go their separate
ways. Shockley had found a new mentor, much more to his liking, Philip
McCord Morse.

Morse was perhaps the biggest influence on Shockley's scientific life.
Although only seven years his senior, Morse served in some ways as the
substitute for the bearded elder who died when Shockley was still a
teenager.

Shockley couldn't have picked a better mentor. Morse was a tall,
handsome man with full curly hair, dark eyebrows and a trimmed moustache on an ovate face – he looked very much the 1930s film star who
specialized in playing the suave protagonist when young, and sometimes
the intoxicated uncle later on. He loved music, good food, drink and
women, not necessarily in that order. In his autobiography, published by
the staid MIT Press, he discreetly describes his active and rewarding sex
life before he married Annabella, the friend of one of his lovers. Morse
dressed well and lived well. He was charming, funny, and extremely
bright, although he freely admitted he lacked the imagination, the
genius, to make the historic intellectual leaps of the great scientist. A
first-rate teacher, two of his protégés won Nobels, Shockley and Richard
Feynman.

Morse was born in Shreveport, Louisiana in 1903. His father, a telephone engineer, was in Shreveport helping wire a new telephone system,
although the family came from Cleveland and that's where Phil Morse
grew up. He could trace his family on his father's side back to Massachusetts in 1660, almost as long as Shockley could. His mother was a
reporter on her father's newspaper in East Liverpool, Ohio.

Morse went to Princeton and immediately came under the influence
of Karl Compton. He learned a great deal. 'Karl believed in the best of
everyone, and one had to oblige him by doing one's best.'

Morse also learned there were more things in life than science,
including music and literature. The rest of his life he made sure he read
at least two books a week that had nothing to do with physics.

'Whatever the subject of conversation, Phil would display... the amazing span of his reading, his knowledge of history and literature, his familiarity with art, his love of music and the theater,' wrote one of his colleagues.[40] Morse also had an uncle who was a newspaper reporter in New York City, and he learned the location of every speakeasy in Manhattan, knowledge he shared with his friends and students.

Morse studied plasma physics at Princeton, the life of atoms in gas, and was tuned in to the amazing discoveries coming out of Germany and Denmark. In 1929, through Compton, Morse got a summer job after graduation at AT&T's Bell Telephone Laboratories, then located on West Street in New York City.

An industrialist, Frank B. Jewett, who had a PhD from Chicago under Millikan, essentially created the lab early in the century. Jewett worked for Western Electric on a nagging problem that pestered the telephone monopoly from the beginning: the inability to make telephone calls over long distances because of the need to repeat or amplify the sound along the wires. At the time, New Yorkers could call no further than Chicago with any clarity. The answer, Jewett thought, lay in vacuum tubes, but the ones being produced then wouldn't do the job. In 1910, Jewett suggested AT&T put together a research team to tackle the problem by hiring 'skilled physicists.'

Until then, AT&T had been relying on Alexander Graham Bell's patents and buying rights from other inventors to keep its monopoly flush. The company's chief engineer, John J. Carty, liked Jewett's idea because it meant AT&T could push the technology and protect its monopoly by acquiring more of its own patents. By 1911, Jewett had assembled a corps of engineers and physicists to attack the long-distance problem, and in 1925, four years before Morse arrived, Bell Telephone Laboratories was born.[16]

Morse worked for Clinton J. Davisson (who would win a Nobel Prize in 1937 for his work on the diffraction of electrons by crystals). Davisson turned him loose on research into the behavior of electrons in a crystal lattice. Morse spent some of his most valued time, however, as a member of the Three-Hours-for-Lunch Club, an *ad hoc* group of Bell Labs boffins who met regularly and leisurely for a wet repast, generally held in local speakeasies. The group included Davisson, his colleague Lester Germer, Davisson's boss Mervin Kelly, and the physicist Walter Brattain. The connections Morse made at Bell Labs during that well-lubricated summer of 1929 were passed on to his favorite students, including Shockley and Fisk.[39]

THE LIGHTNESS OF BEING 41

While we're on the subject of the Three-Hours-for-Lunch Club, Germer's legacy was particularly interesting. Besides being a fine scientist, he was an accomplished rock climber, scaling steep cliffs for the fun of it, rare in the 1930s. He introduced Morse to rock climbing at an encampment in New Hampshire. Morse passed on his passion to Bill Shockley. Germer died of a heart attack while climbing the Shawangunk Ridge in New York State, a climb Shockley later made a point of mastering.

Compton brought Morse to MIT in 1931 when he took over as president.

Morse wrote what became the standard text on acoustics and, with Edward Condon, the first English text on quantum theory. Morse devised several techniques that Shockley used for his dissertation and for his published papers, but refused to be given credit.

Shockley was three months from graduation when he received an offer for a temporary teaching job at Yale. The lectureship was one turned down by Fisk for a better offer from the University of North Carolina. Slater recommended Shockley. It was 1936 and America was still locked in the Great Depression. Any job offer was a gift, especially in science, and Shockley, who now had a wife and daughter, had little choice. The same day, Mervin Kelly, he of Bell Labs' Three-Hours-For-Lunch Club, visited MIT to interview possible candidates. The labs were now ready to hire after suffering a long freeze. Shockley had met Kelly in New York once, probably through Morse. When Morse told him about the Yale offer, Kelly, knowing Morse's high opinion of his grad student, called New York for authorization to make Shockley an offer on the spot.[41]

Shockley graduated on 26 June 1936 at Orchestra Hall in Boston, with May in the audience. Slater was heard to say that with Shockley's graduation, he – Slater – had lost his best teacher.

The graduating class held a party at Lochober's restaurant in Boston that night. No one thought to invite Slater.[20]

PART II
Hubris: war and the transistor

CHAPTER 3
'Of a highly explosive character'

One day in the early 1930s, at the new facilities that AT&T had built for its Bell Telephone Laboratories in Murray Hill, New Jersey, Karl Darrow, a man with no discernible sense of humor, stood at a podium and began to lecture.

Darrow's talk was part of a program begun by William Shockley and his boss, Mervin Kelly, shortly after Shockley arrived from MIT to keep the scientists at the laboratories up to date on the latest developments in solid state physics. It was audacious of Shockley, newly arrived, to organize such a program, but his efforts were welcomed and the lectures well received. Solid state physics had become complicated stuff, and Darrow had brought a series of slides to illustrate his lecture on the screen behind him. The large auditorium was packed. To this day, no one remembers what Darrow said, though they will never forget the lecture.

Shortly after Darrow began, a small wind-up mechanical duck waddled and quacked from off stage right and continued across the stage behind him.

The muffled giggles began as the men in the audience tried to watch the duck and keep from disrupting Darrow, but they could not hold back for long. Darrow could not see the toy, but he could hear the quacking and the laughter, and saw the audience's attention disintegrate. He finally stopped talking and turned around when the duck was behind him. The duck kept going. Darrow stared silently with a totally blank face, following the duck as it waddled and quacked itself offstage and the audience fell apart.

Saying nothing, Darrow walked out of the auditorium, his face a stern mask.[*]

---

[*] Darrow had more class than his supporters suspected; he and Shockley became friends.

Some in the audience, appalled at the lack of manners, stunned that the Bell Telephone Laboratories could treat such a redoubtable gentleman with disrespect, stomped from the room in anger. The rest of the audience – the majority – was in pandemonium, screaming and weeping with laughter. Some of the men had even fallen off their chairs.

People at the lab spoke about nothing else for weeks. The only thing everyone in the labs could agree on that morning was who was responsible for the duck: Bill Shockley.[1,2]

—<

With Alison in tow, the Shockleys regularly took the train from Boston to New York to find a place near Bell Labs' facility in lower Manhattan. Their quest was fruitless and discouraging for most of the summer. Even with a long list of available apartments, they found none they liked and no one answered their knocks or phone calls at many of the addresses. Part of the problem was finding a nursery school for Alison. They checked out several and found they would cost $240 a year or $30 a month at the least. That expense eliminated a number of apartments.[3]

Most of their friends were in Greenwich Village and Jean was particularly interested in living near there. It wasn't until shortly after Labor Day and dozens of trips to the city that they found what they wanted.

The brownstone apartment at 258 West 17th Street had two freshly painted bedrooms. The house was attractive and well kept; the kitchen and bath tiled and as modern as any apartment they visited; and high ceilings and tall windows let in a warm stream of diffuse light. Visitors entered into a central hall. The living room was on the left and the bedrooms were in the back of the apartment, behind a dining area. The only drawback was that the second bedroom, Alison's, connected to theirs, reducing their privacy.

The neighborhood was an unattractive commercial district just off the Village. That meant cheap rent; the apartment cost $60 a month, which they could afford with some effort. Additionally, it was within walking distance of Shockley's new office, on the 12th floor of the Graybar-Vanick Building at 463 West Street. Jean was enormously pleased. May grumbled that the Village was dirty and they ought to find someplace better. Their apartment, Jean admitted, was on the ground floor, which made it more susceptible to city grime, but: 'We moved six times in our three years in Cambridge. I am just sick enough of the process to want to stay put for a while and give a place a chance to grow on us – especially as we are conveniently located and very well pleased with our apartment in itself.' Unlike her mother-in-law, she wasn't a supporter of relentless searching for someplace better. She now had a daughter – a family – and she wanted to settle in.[4] End of discussion.

Jean found work at the National Board of the YWCA to supplement their income. Alison went to the Bank Street Nursery School. Shockley could walk hand-in-hand with his daughter down West 17th Street, drop her off at school and continue on to the laboratory.

Shockley was making $310 a month and enjoying life in the big city, although funds were short, as he occasionally noted to May. 'I had hoped that we would be able to live on the salary this year, but as things work out, it limits us pretty much as regards to entertainment etc.,' he wrote. 'In fact, since the first week we have been here, Jean and I have been out to dinner to a restaurant by ourselves by way of entertainment only once.'[5] May sent a check. Shockley and Jean bought concert tickets for alternate Thursday evenings for $10 the season.[6]

Shockley kept to his strenuous physical regimen. He exercised regularly with barbells and found that all but one of his suits grew too tight. 'I wear a coat only a small fraction of the time, whereas I wear a suit all day,' he wrote his mother after she sent him a Christmas check for new clothing.[7] He used the money to buy another suit.

He reveled in being in peak physical condition and continued to use sunlamps. He sometimes entered a room first by grabbing the top of a door, chinning himself up and then back flipping into the room, mostly because he could,[2] which some at Bell Labs found disconcerting.

Shockley's only physical concern was that he was growing bald. Jean seemed far more dismayed than was he. During the early years in the city, he seemed to enjoy the regularity of life. He came home after work, exercised, took a shower, read and listened to the radio in the corner of the living room, mostly to concerts. He and Jean seemed pleased at the wide range of radio programs in New York. Dinner came at seven or seven-thirty. He read and worked until bedtime.

Religion was not a presence in the home. Jean, the daughter of a would-be missionary, read the Bible a lot more than did Shockley, who had none in his home as a child. One day Alison came home from school singing a garbled version of 'Jesus Loves Me.' Shockley suggested they teach her 'Thor could lick the whole Holy Family with Moses thrown in.'[8] Shockley attended meetings of the Ethical Culture Society (an unreligious, free-thought community) in New York several times but did not go to church.

Shockley stood duty at parties for Alison's little friends, bringing his bag of magic tricks. After a camping trip, he grew a moustache and beard, and for one performance, soon after, he dressed with a colorful kimono draped like a turban over his head, looking like a dire shadow from the Arabian Nights. The kids loved it. The beard came off a week later.

At the labs, Shockley was assigned to work under Davisson. Shockley was supposed to study valves, devices that controlled the flow of electricity – the key to telephone switching – but his lab superiors made it clear they would not hold a tight leash on his work. The job had a steady nine-to-five schedule at first, with, Shockley admitted, nothing resembling pressure. He eventually had to work the long hours common among the veteran researchers, but for the first half-year a company orientation controlled his time.[9]

When Davisson went on a long trip via Sweden the next year to pick up his Nobel, Shockley moved into his office (sharing it with Darrow) and used his desk. He could avoid distraction there as it was out of the way. He was a young star at the labs, and visitors often dropped in to chat, sometimes on topics he knew nothing about. His productivity increased considerably once he moved into Davisson's realm.[10]

He published eight articles in scientific journals that first year, including one he began at MIT for *Physical Review*. More important, he dived into the great participation sport at Bell Labs: patents. Within a year he had filed his first one, for a device that keeps a stream of electrons focused through a screen.[6,10] Earning patents meant winning your stripes at Bell Labs.

Several of his friends joined him at Bell, including Fisk, who had left North Carolina. Shockley was happy to see him. Fisk had been an ally in mischief and the two men – at least at this stage – enjoyed their intellectual competition.

Another new compatriot was John Pierce, who arrived at Bell from Caltech with a degree in electrical engineering about the same time as Shockley. Pierce found himself in the vacuum tube laboratory where he met Shockley, then meandering from lab to lab.

They had similar interests: electron multipliers, devices used to increase or enhance an electrical effect. 'If you strike a surface of the right consistency with one electron, about 100 volts,' Pierce explained, 'you'll get four electrons out' using a multiplier. They demonstrated this on a four-by-eight rubber sheet using ball bearings to simulate electrons.

This had some immediate practical implications: the multipliers could easily be used for recording and playing back the sound on films. Western Electric, the AT&T subsidiary that half owned Bell Labs, supplied much of the sound recording equipment to the burgeoning film industry and they wanted to keep up with the technology.

Shockley and Pierce published a paper on their work; then Shockley went on to other things. The two remained friends, dining at each other's homes, Pierce awed by Shockley's physical condition. 'He could walk on his hands,' Pierce remembered later. 'He practiced on the roof of his house.' The two also went to cultural talks in New York together, and to lectures on contemporary music. 'He knew a lot of things very deeply, and other things that were very common, he had never heard of,' Pierce recalled.[12]

When construction began on Bell Labs's new facilities in northern New Jersey in 1937, Shockley knew that eventually he was going to be transferred, but Jean discovered that the schools in New Jersey would not take Alison yet because of her late-in-the-year birthday. Private schools still were too expensive. A real estate agent in New Jersey assured them they had a 'steal' with their West 17th Street apartment so they decided to stay put another year.[13] Shockley could commute by train when his job moved to New Jersey.

That year, the family began a ritual, summer vacation at Lake George in Upstate New York. They found a virtually uninhabited island and camped for a week or two almost every summer. The island had a little swimming cove facing south where Alison could wade. The bottom gradually dropped into deeper water, perfect for adult swimmers. Because it was enclosed, the water in the cove remained sun-warmed no matter what the temperature in the rest of the lake. For a decade thereafter, the island became one of Alison's fondest childhood memories, and the place the family was happiest.

Bill introduced Jean to climbing, using a mountain they called the Elephant, across from the island, as her school. She soon became reasonably adept, though hardly as daring as her husband, who would eventually take up rock cliffs and the Alps. One day in August 1937, after leaving Alison with friends, they canoed across the lake to the Elephant and instead of trying to find a path up the rounded peak, climbed straight up it, Shockley blazing the trail for her, snapping off small

Figure 9 The Shockleys at Lake George, just after the war.

branches. The country was wild and tumbled with occasional small cliffs to be skirted, thickets, rocks, and small meadows interspersed in the deciduous forest. They would climb a way, stop to look back across the lake at their island to orient themselves, then continue again.

Shockley also continued his fascination with caves, finding a number of interesting examples near Schenectady, often dragging Seitz along

Figure 10 Shockley climbing in the Adirondacks.

with him. One cluster of caves had swarms of bats, one of which Shockley took back to show Alison. Jean was not amused.

One facet of her husband's personality seemed to bemuse Jean – Shockley seemed to have odd notions about race. In September 1939, she wrote May:

> Yesterday I carried out a little strategy in interracial education, which I have long contemplated. Bill is fascinated by the South Seas and hopes to be able to go in search of some [islands], one day – if not in the South Pacific, perhaps in tropical Atlantic waters. One of the stenographers here – a colored girl with a wonderful name, Pocahontas Foster – spent seven weeks in Martinique last summer with her husband. So I got Bill to come up and have lunch with Pocahontas and me here at the Y so that she could show us her pictures of Martinique and tell us about the trip.
>
> It was really fun, and I was surprised that Martha Eddy, my office manager with whom I usually lunch, came along too when I invited her. So I rounded up two people who are half convinced that the Negro is really an inferior race, and who, like me, don't know any Negroes at all well, and we had quite a party. Pocahontas is well adjusted enough to be able to talk about race with perfect freedom, and I'm sure she didn't feel she was being exploited. She is a very cute youngster with a tendency towards elegance in her vocabulary, which gave no impression of affectation, however, and what she had to say of Martinique was very interesting. One reason she and her husband decided to go there was that the color line is not drawn with devastating severity as in the British-administered islands, and it was more interesting to go to a place where a different language is spoken. She speaks French very well, and apparently enjoyed practicing and improving it. She says that the population includes all shades from white to black, and that the official class is more creole than white.
>
> I was naturally pleased to have Bill say, when I asked him if he thought she was below the average white stenographer, that she seemed considerably above![15]

The origin of Shockley's concept of race is unknown and unknowable. It was not a sudden vision in later life, but one held throughout his life. He was 29 when Jean wrote that letter.

Undoubtedly, he absorbed some of his racial theories from May and William, but their beliefs on race were hardly unusual for their time, when America was totally segregated and African-Americans were only two generations from slavery, and it did not seem of great importance in their lives: they mentioned it only in passing. Shockley himself had virtually no contact with African-Americans. There was one in his class at Caltech, Grant Delbert Venerable, Jr, who became a moderately successful business executive. Venerable was likely the only African-American with whom Shockley had any prolonged contact, but if he did, neither man ever mentioned it. It is not likely there were many African-Americans – if any – at MIT, and Bell Labs had none above the menial level.

—<

Two of Jean and Bill's companions on their trips to Lake George were Walter Brattain and his wife, Keren, a chemist.

Brattain, eight years older than Shockley, came from the Pacific Northwest, one of a long line of young men, including Linus Pauling, who grew up in the passing shadow of the frontier and went on to become distinguished scientists. He was born in Amoy, China, on 10 February 1902, where his father taught at the Ting-Wen Institute for Boys. The family returned to the Brattains' native Washington State when Walter was a year-and-a-half old. They lived in Spokane until he was nine, where his father worked as a stockbroker, then on a ranch near Tonasket, in the Okanagon Valley of eastern Washington.

Both Brattain's parents came from pioneer stock. His paternal grandfather made the trek on the Oregon Trail in 1852 as a boy of 16. Walter's grandmother, born in New Brunswick, Canada, crossed the plains to the Northwest when she was four with her Scottish-born parents. His maternal grandfather, a native of Stuttgart, journeyed to San Francisco in 1854, bringing his bride over later. They settled in Pomeroy, Washington, where they opened a flourmill.

Walter was the oldest of five, including two sisters who died as children, a surviving sister Mari and a brother Robert, who also became scientists (a graduate student with Seitz at Princeton). Both his mother and father went to Whitman College in Walla Walla. The family was highly literate, taking time after lunch so the parents could read books aloud to their children.

The Brattain brothers were cowboys, riding the range alone or together with their cattle from the age of six, usually armed and wearing leather chaps for protection against rattlesnakes and thorns. The area swarmed with rattlers; they shot one almost every day. At the age of 14, Walter could spend weeks alone in the hills with the cattle and two or three horses.[16]

He followed his parents to Whitman, a place he learned to love. There, like so many great scientists, he ran into great teachers, in his case Benjamin Brown (physics) and Walter A. Bratton (math). Both had taught Brattain's parents.[17] You couldn't talk to Brattain at any length about his life without his mentioning his gratitude and devotion to these men, especially Brown. Plunked in a 500-student college in the middle of a distant desert, the two men specialized in turning unsophisticated, poorly educated kids from nearby farms and ranches into fine scientists and engineers year after year. The year Brattain graduated, he was one of four Whitman physics majors. All four ranked in the top ten in a national contest for a Harvard fellowship, including the winner.

Brattain went on to the University of Oregon for his masters in physics, and earned a scholarship at the University of Minnesota to study under John T. Tate and J. H. Van Vleck for his doctorate.

Like Shockley's time at Caltech, Brattain's stint at Minnesota came at the cusp of change in the study of physics. Tate had gone to Germany to study because that's what you did then if you wanted to keep up with the field. By the time Brattain got to Minnesota in the mid-1920s, the tide had shifted and America was catching up. 'My age comes almost at the dividing line when it ceased to be necessary for an American physicist to go to Germany in order to finish up his degree,' he said.[18]

Btattain was about five-nine, left-handed, slim with blue eyes and blond hair.

He had received the first inkling of the revolution in physics at Whitman: references to Planck's quanta in Brown's class. Van Vleck's course at Minnesota was purely theoretical and, of necessity, highly immersed in the new quantum concepts. That meant no text; how could you write and print a text with things turning upside down every week? Even shockwave riders like Van Vleck had to hustle to keep their classes current.

Graduating from Minnesota in 1928, Brattain applied for jobs at the National Bureau of Standards in Washington and the Bell Labs. The bureau offered him a job; Bell dawdled with his application. He accepted the Bureau job despite his fears that the position, which involved research on oscillators, would turn him into a radio engineer, not a scientist. Several months after arriving, he was selected to give a tour to a visiting Bell Labs executive. The man was impressed enough to tell Brattain to call him if he ever wanted to work at Bell. Brattain called, and joined the Bell Labs staff in 1929.

By the time Shockley reached the labs in 1936, Brattain had become a first-class experimentalist, the one you went to when you had a difficult physics problem and you needed someone to design the experiment and run it.

His principal interest at the time was copper oxide rectifiers. Copper oxide is a complex semiconductor, a substance that acts somewhere between an insulator and a conductor when confronted with electricity; a rectifier converts alternating current into direct current by allowing more electric current to flow in one direction than in the other. Diodes are rectifiers; so too the 'cat's whisker' in early radios, the most important example of varying conductivity in a semiconductor until the invention of the transistor. The 'cat's whisker' was to be of great use in radar during the coming war.[19]

All this interested the labs because its telephone engineers needed efficient rectifiers and were trying to use copper oxide in their circuits. Among Brattain's assignments was one to try to find out where in a copper oxide crystal the rectification occurred so that he could study the effect.

Brattain met Shockley shortly after he arrived. They were fascinated by the same things and quickly became friends.[20]

'He was of the next generation, a generation who had gone through college and graduate school saturated in the new quantum mechanics, not one who had to pick it up on his way,' Brattain said years later. '[He] was particularly adept at applying the quantum mechanics to particular problems. He used to discuss and explain [it] to the older physicists who came before me, who had not even grown up in this period, for whom quantum mechanics was a completely foreign thing, something they could hardly understand.'[18]

This insight was crucial. 'It's perfectly clear that people like Slater, like Fred Seitz, Wigner... and a few others were beginning to apply quantum mechanical concepts to the study of solids,' Fisk said later, 'and there seemed no doubt that in due course there would develop a far better understanding of solids as the whole subject advanced.'[11]

Shockley organized an informal study group in New Jersey that included Fisk, the metallurgist Foster Nix, Pierce, physicists Alan

Holden and Charles Townes (who would later help invent the laser), Brattain and several others. The men met twice a week to talk about the possibilities in the new physics. Participants took turns preparing material for the 4:30 p.m. sessions, sometimes bringing in new texts, including Tolman's book on statistical mechanics. The only problem, Shockley felt, was discouraging 'some guy who's just not well enough qualified to be there, [who] is going to be a dead weight on this activity.'[21]

The lab's research director, Mervin Kelly, believed a day was coming when the mechanical switching exchanges that served as the heart of the telephone network would be replaced by electronic alternatives. He had no idea in the late 1930s exactly how. Shockley believed Kelly was right and thought the answer lay in solid state physics, the physics of electrons flowing through solid materials – hence the study group.

Shockley's first notion toward this end was a device that had carbon contacts in a quartz crystal. He hoped this would produce an amplifier, something that increases the strength of an electrical signal. Others at the lab had tried this before and failed.

In December 1938, Shockley went to Brattain and suggested that if they made a copper oxide rectifier in just the right way, maybe it would act as an amplifier. Shockley's idea this time was to insert a grid into the layer of oxide on the copper to control the current, like a gate or valve.

Shockley made an entry in his laboratory notes, as required by Bell Labs procedures, on 29 December 1939, depicting just such a device. 'It has occurred to me that an amplifier using semiconductors rather than vacuum is in principle entirely possible,' he wrote.[22] He didn't bother to have the entry witnessed for two months as procedures required. This kind of entry was called a 'disclosure,' a step toward a potential patent. On 27 February, one J. A. Becker endorsed the entry with 'read-understood,' as was customary.[23]

Shockley knew that if anyone in the labs could create such a device, it would be Brattain; no one could design and construct an experiment like him. Brattain listened to the idea – he liked Shockley – but he pointed out that he and Becker had already tried something similar and it hadn't worked. But because of Shockley's eagerness, Brattain couldn't turn him down.

'Bill,' he said, 'if it's so damned important, if you tell me how you want it made, if it's possible, we'll make it that way. We'll try it.' He made several units to Shockley's specifications. The devices produced 'nil,' Brattain reported. The results were not encouraging, Shockley admitted. Nonetheless, the theory appeared sound, and on 29 February 1940, two days after Becker endorsed Shockley's first entry, Brattain (with some amusement) and $W = Shockley$ signed a laboratory notebook entry that described an improved device that theoretically could have produced a semiconductor amplifier.

Others at Bell Labs nibbled around the edges of this problem at the same time. One group tried to produce a similar device with sodium fluoride and failed. Yet another, led by Russell B. Ohl, produced crystals of silicon, a well-known semiconductor that could act as a rectifier. They accidentally manufactured one 'ingot' that rectified in one direction from one half, and in the opposite direction from the other half, a phenomenon none of them could explain. The rectification, they found, switched direction in a very narrow boundary of atoms within the crystal. In the winter of 1940, Ohl and his team demonstrated what they had found in Kelly's office. Wires led from the crystal to a voltmeter. Black as coal, the crystal measured 0.5 centimeters by 1.5 centimeters. Kelly and Brattain watched as the men shone a flashlight at the silicon. The crystal acted like a photocell, turning the photons from the flashlight into electrical energy. With astonishment, however, they watched the meter jump almost half a volt. The silicon multiplied the energy ten times more than they expected for a crude hunk of crystal without polished surfaces. Brattain said he thought someone was pulling his leg.[24]

The boundary within the crystal where the rectification switched directions is called a junction and the whole piece would now be described as a p-n (positive-negative) junction. While others had tried this and produced ambiguous results, this was 'the first one that I ever observed that was clear-cut,' Brattain said.

In the summer of 1939, the Shockleys decided to move. While the Murray Hill facilities were under construction – they weren't ready until 1941 – the lab used temporary headquarters in Whippany, west of Newark. Trains ran between that area and New York City and Shockley could commute, but Alison now was ready for school.

They found an eccentric wooden house hand-built by a one-time ship's carpenter named Otto Karlson in the woods near the little village of Gillette. A family that rented it for the summer had just vacated the house. The carpenter gave it to the Shockleys for $10 a month less than the rent for their New York apartment. It had a fireplace and oil heat operating on a thermostat, unusual for that time. There were four bedrooms upstairs and two baths; a living room, dining room, kitchen, breakfast room, a large porch and a small bedroom downstairs that Shockley could use as an office. Fine – except the house wasn't finished.

All the walls were white plaster, some of the plaster painted over with white paint. Shockley was grateful Karlson hadn't finished the walls because, he said, he had 'absolutely no taste at all and the wallpaper he would have selected would probably be unbearable.' One of the bathrooms was painted 'violent lavender,' Alison remembered, and the other had black tiles. They abutted each other, which made the plumbing easier for Karlson to install, but all the plumbing never quite worked at the same time.

The Gillette house, on Long Hill Road, was relatively isolated, vastly different from West 17th Street. There was another house to the west, four to the east, all built at intervals of 300 feet; beyond that, nothing but woods for half a mile. 'We get quite a clear view north,' Shockley wrote May. 'When the leaves come off the trees to the south this winter, we shall have a view that way.'[25]

'I can remember going out into the woods with him on weekends and observing nature,' Alison remembered years later. 'One of the things that we did – it became a project after a while – was to bring home snakes, many different ones. In those days, I could and he could identify them. None were dangerous; I never came on a copperhead. Mother stepped on a copperhead once, on its head. She was lucky.' They put the snakes in the bathtub of whichever bathroom was useless at the time, and, as a scientific experiment, tested them to see if they were constrictors. The next day they let the snakes loose.

Shockley set up a firing range in the basement with a pile of sand, popping away with a .22 long rifle.

Jean got out binoculars and sat by the kitchen window, watching the birds. She could go out on the porch to feed them during the winter.

Shockley got up at six every morning to reach the labs back in New York by 8:45. First, he shaved and combed his hair, dressed in a sweat shirt, blue pants and tennis shoes, and went running for half a mile, sometimes taking their new cat, Belle, with him (he wrote May), although it was not clear how that worked. He returned, lifted weights, and then took a warm shower that ended with a splash of cold water. He sat in the kitchen for breakfast at 6:55 and made it to the 7:55 train.[26]

Jean drove him to the train station every morning, unless he decided to walk, a hike of only about 10 minutes once he and Jean cut steps in the hill directly opposite the house. Shockley then could climb the little hill, walk swiftly through the woods and leap down to the road leading to Gillette's Lackawanna Railroad station, cutting 15 or 20 minutes off the walk.

When the lab finally moved out of New York, Shockley was assigned to the radio laboratory in Whippany until the new facility in Murray Hill opened. Whippany was about a 25 minute drive. He bought a used 1932 LaSalle for the commute.

Jean signed up for a graduate course in educational psychology and one in teaching high school English at Montclair State Teachers College, which would allow her to teach in New Jersey. She dropped Alison off at a nursery school in South Orange on the way to Montclair, or with a neighbor who could watch her. The problem was finding a job. May offered to fly Jean and Alison west for the summer of 1939; Jean turned her down lest a job opportunity knock, which it did not.[27]

They loved the Gillette house.

The outside world took this opportunity to disrupt Shockley's research. It began one of the most bizarre and least known chapters in the history of the Nuclear Age.

Around Christmas, 1938, the Austrian scientist Lise Meitner prepared a letter to the British scientific journal *Nature*. Meitner had been working with the physicist Otto Hahn studying the uranium atom until Hitler annexed Austria. Meitner, who was Jewish, immediately fled to Stockholm to continue her work, dealing with Hahn and his associate Fritz Strassman by telephone and mail. The letter announced a stunning idea: It was possible to split the nucleus of the uranium atom by slamming a neutron into it.

Her nephew, Otto Frisch, also a physicist, called the process 'fission.' It wasn't even hard to do, she reported. The result of such a collision would be an unusual amount of energy and two or three additional neutrons, useful for bombarding other nuclei.

Frisch took the draft of her letter to Bohr in Copenhagen, then just packing for a trip to the US in February 1939. Bohr, instantly recognizing the importance of the discovery (which he quickly confirmed in the laboratory), decided to bring the news to the American scientists even before publication. He packed the letter in his luggage and sailed to New York.

Shockley, Brattain and Fisk were in the audience at one seminar Bohr gave at Columbia University. They took the message back with them to the labs, and the possibilities caused considerable unease and interest.[28]

(The implications were hardly lost on many outside the labs, either. Everyone was aware that within Meitner's fission lay the potential for a weapon of horrific power. Refugee scientists Leo Szilard at Columbia and Eugene Wigner – Fred Seitz's collaborator at Princeton – tracked down Albert Einstein on vacation and urged him to bring the matter to the attention of President Roosevelt. Einstein did, in the celebrated letter that led to the Manhattan Project. Among those also getting to work immediately to test the potential was Enrico Fermi, then at Columbia.)

In May 1940, Kelly asked Shockley and Fisk to answer the following question: 'Can nuclear energy be made available by the fission process?'

The president of Bell Labs, Frank Jewett, at the time also served as president of the National Academy of Sciences. The academy's National Research Council, along with the National Bureau of Standards, were marshaling the scientific troops across the country to work on the fission problem, fearing that Hitler's scientists, who included Heisenberg, were doing the same. The labs gave Shockley and Fisk their own room and all the equipment they needed. The two did not intend to build a reactor or a bomb; their quest was simply to find out how to design a device that could produce nuclear energy.

A few days after getting the assignment, Shockley had a eureka moment standing in the shower in Gillette. The solution was to slow down the multiplication of flying neutrons so that the chain reaction could be controlled while at the same time keeping the reaction going.

'Well,' Shockley told himself, 'if you put the uranium (^{235}U) in chunks, separated lumps or something, the neutrons might be able to slow down... and not get captured, and then be able to hit the ^{235}U. So Fisk and I settled down and did calculations on what optimum dimensions would be, for layers and cylinders and spheres of non-enriched uranium in graphite and water,' Shockley said.[21] (His insight wasn't unique, he readily admitted; at least three or four people figured this out simultaneously around the country, including Fermi who produced the first chain reaction in 1942.)

Voluntarily, without any pressure from the government, scientists working on the fission problem decided not to publish any of their findings lest they inadvertently help the Germans. Shockley wrote May about the work, saying only it involved power from uranium and advising, 'this should not be repeated, of course.' He added that all his future plans depended on what they found. It might be that the lab wanted him to go west to visit some laboratories there.[29]

It took Shockley and Fisk two months to solve the problem. They reported to Kelly that you could indeed generate energy with a nuclear reactor – and then some.

They suggested separating layers of uranium (the 'U layers') by using a paraffin containing hydrogen or by using graphite to slow the neutrons. One layer would be the rare ^{235}U and the next, its far more common isotope, ^{238}U, and so on, alternating.

Besides being useful as a power source, such a reactor would have other uses:

- As an explosive.
- As a source of 'artificial' radioactivity. 'If the chain reaction could be made to go, large quantities of radioactive material would be created at, we believe, a controllable ratio.'
- As a neutron source.

'The other uses,' the men agreed, 'are probably more interesting.'

'As a source of power, uranium has many obvious possibilities such as a power supply for long-range flying or long-range cruising underwater or for power in isolated places,' Fisk concluded in his laboratory notebook.[30]

They added one chapter to their report called 'Rather Vague Considerations Concerning the Explosive Aspect of the Chain Reaction.'

'The energies involved in this state of affairs – U layers at several million degrees – would appear to be of a highly explosive character,' they wrote. The energy would be concentrated in thin layers and blow up rather than just get hot.[31]

Shockley and Fisk added that the most important finding was that tons of uranium would not be necessary as first speculated. A few hundred pounds would suffice. Nor would large-scale methods for isolating the isotopes of the rare element be required.

The moment they sent their report to Kelly and to Lyman Briggs at the National Bureau of Standards, the lab shut them down. Perhaps the lab merely felt the work obviously had nothing to do with telephony. More likely, the government slammed a secrecy lid on the project.

The report wasn't entirely squelched; it was believed to have influenced nuclear work in Britain and Canada, but it had no impact in the US. The reason, Fisk later wrote, was government embarrassment that two men at Bell Labs, not involved in the huge federally funded projects, could solve most of the problems in two months.[11]

The two men applied for a patent on the Fisk–Shockley reactor. 'I know for a fact that all these applications went into the secrecy file' along with all the other patent applications for similar work, said Brattain. After the war, when they opened the file to see who should get credit; the Fisk–Shockley application was dated the earliest, to the apparent chagrin of the government. The Atomic Energy Commission and the patent office 'connived to try to throw the application out on every conceivable ground. The book was thrown at it,' Fisk said.[11] The government wanted credit to go elsewhere, perhaps to a public project as opposed to the Bell Labs' private efforts. In the end, the two men assigned the patent to AT&T's Western Electric division, which promptly dropped it.

'Actually, the work was very largely Bill Shockley's genius,' Fisk said years later. 'The only thing that bothers me is that he's never had any credit for it. As far as I'm concerned, I was riding along, whipping up the horses.'[11]

The work wasn't completely forgotten. Six years later, Buckley, then president of the labs, got a letter from Briggs, chairman of the Department of Commerce's Uranium Committee, thanking Shockley and Fisk. 'Their report proved to be a remarkably accurate forecast of what might be accomplished when we consider the paucity of information available at the time.' Briggs noted that the two men were not part of the nuclear research establishment, which made their work even more valuable, acting as a kind of external confirmation. He said that Vannevar Bush, the *eminence grise* of America's early 20th century scientific establishment, told a scientific organization early in the war he wanted to hear every suggestion. 'If they are good, they will be used but you probably won't hear about it until after the war is over,' Bush said.

'I record now my grateful appreciation of Fisk and Shockley's contribution,' Briggs wrote.[32]

On the other hand, in 1948, Eugene Wigner, who worked on the Manhattan Project, wrote to Fisk, by then head of the Atomic Energy

Commission, congratulating him on a report he and Shockley sent him. 'I am a little in the dark as to why we had no access to this report,' Wigner wrote.[33] The government had even kept it from its own scientists.

This wasn't Shockley's last endeavor in the uses of nuclear energy or its potential for destructiveness.

—<

Jean, now commuting to her job at the Y from New Jersey, saw the change at the Port of New York as she rode the Hoboken ferry across the Hudson. In September of 1939, after war broke out in Europe, she noticed far fewer ocean liners were arriving. The British registry *Monarch of Bermuda* steamed up-river, her whole hull and superstructure now a battleship grey. Even her name was slightly smudged. Two Cunarders in the White Star docks were being painted similar shades.

In the Holland-American docks the *Veendam* too had a different paint scheme: The name VEENDAM-HOLLAND was painted in huge letters on her side with a large red, white and blue striped rectangle next to it, to show the German submarines that she came from a neutral nation and her destination was a then-neutral US.

Within two years the US would be at war, and scientists like Shockley would do battle – with slide rules, pencils and graph paper.

For Bill Shockley, it would be among the best of times – and the beginning of the worst.

CHAPTER 4

'I hope you have better luck in the future'

Five years after it was born in the back seat of May's Buick, Shockley's marriage began to fall apart.

He and Jean began fighting over household affairs. They fought over money. They fought over Alison's upbringing. Jean's letters to May slowed to a trickle, mostly reflecting the banality of her daily life. Shockley claimed he was too busy to write. May had visited several times in Cambridge and New York to babysit Alison, but with the war coming on, travel became harder. When she did come, she and her son often fought, their personalities too similar for peaceful coexistence.

Jean managed to get a substitute teaching job at Plainfield High School, which gave her some distractions.

One day Shockley asked Jean to take Alison in for an IQ test, out of curiosity, he said. She refused 'for fear of taking it too seriously one way or the other.'[36]

In the summer of 1941, Karlson, their Gillette house landlord, tried to raise their rent to $80. Rather than paying that much, the Shockleys moved to a house in the middle of nearby Madison. The new house, at 45 Maple Avenue, near Drew University, was considerably smaller, two storied, with a columned porch and a driveway on the side leading to a one-car garage and a small backyard. Shockley walked to the Madison train station a long block-and-a-half away and Jean could amble to the shopping area behind the station. The house had three rooms downstairs, a living room, dining room and kitchen, all smaller than the rooms in Gillette, four rooms upstairs and a bath, giving them a guest bedroom that could serve as a sewing room for Jean, and a study for Shockley.[36] Shockley said the house's small size fit their furniture better.

Shockley began traveling more, which probably exacerbated his growing loneliness. When he returned home he still spent as much time as he could with his daughter. The problems he was having with Jean did not appear to alienate him from Alison. They read the *Oz* books together, much to Alison's delight. She roamed the house dressed as the Ozma of Oz with two bows in her hair instead of poppies. Belle, their cat, played the Cowardly Lion. Jean worried that the stories were too advanced for Alison, but the seven-year-old appeared to love them and Shockley clearly loved reading them to her.[34]

On one of her birthdays, he took the train up from Washington to attend her party, performed a series of well-received magic tricks, got back on the train and returned to Washington.[8]

Despite the increasing pressures in their marriage, Jean became pregnant around Christmas time in 1941.

Shockley went off to war a few months later.

One January day in 1941, Phil Morse, Shockley's mentor at MIT, sat on the edge of a bed in a Washington hotel listening to an excited Navy commander, E. C. Craig, describe the problems the British were having with German acoustic mines. The Germans began dropping the acoustic mines from planes into the seaways, terrifying the British convoys, the embattled island's arteries to the world. Acoustic mines explode at the sound of nearby ships and all ships make noise. As Craig talked, his voice grew louder until an embarrassed Navy lieutenant in the room next door had to knock to suggest the officer be more discreet. Craig asked Morse if he could help find a solution.

Morse, like many American scientists, desperately wanted to be useful during the coming war. 'By that time few of us could think of carrying on as usual; we were ready for someone to tell us what to do,' he wrote.[37] Morse already had volunteered for the Radiation Laboratory (RadLab) at MIT, created by the new National Defense Research Committee, itself a creation of MIT's Vannevar Bush. Bush was appalled at how underutilized scientists had been during the First World War, and convinced President Roosevelt to set up NDRC so that wouldn't happen again in the coming war. Morse found his skills useful to RadLab: what he knew about acoustics had great application for microwave radar.

Ultimately, the solution came down to creating a noise loud enough to trigger the mines from mine sweepers running ahead of the convoys. Almost by accident, someone at MIT thought of dragging two pipes through the water so the turbulence could create underwater noise.

They guessed at the length and diameter of the pipes, and how far apart they needed to be and they were amazed when the parallel pipes produced just the right clamor. Ships trailing them through the water in front of the convoys could sweep the sea-lane of acoustic mines.

The lesson – that you could put scientists to good use even on seemingly mundane problems – seemed lost on the Navy, but not on Bush and Morse. They hired John Tate, Brattain's mentor from Minnesota, to act as liaison with the Navy. At first the Navy resisted Tate's overtures; the brass, especially Admiral Ernest J. King, commander-in-chief and chief of naval operations, had no intention of giving a bunch of civilians access to classified information even if they were apparently loyal PhDs from Harvard and MIT.

In March 1942, Morse met in Boston with another naval officer, Capt. Wilder Baker. Baker had just formed the Anti-Submarine Warfare Unit and he needed help. With the US now three months into the war, German U-boats had been turned loose on the US East Coast with the aim of cutting off the convoys before they even crossed the ocean. Baker and Morse had a mutual friend, the British physicist Patrick M. S. Blackett, who had successfully put a team of British scientists 'in the field' to improve the radar his country needed to defend itself against the Luftwaffe's bombing raids. Baker asked Tate to help him create a similar program. He also went to Admiral King's office and squeezed out reluctant permission for Tate's people to see secret documents.

Morse began recruiting for the Anti-Submarine Warfare Operations Group (ASWORG) in the spring of 1942. He raided Caltech, Princeton, MIT, Harvard and Berkeley, sometimes stealing scientists out of the clutches of other units forming across the country. Shockley got one of the first calls. Morse asked him to serve as director of research with a roving commission – Shockley could go where he wanted, do what he wanted, when he wanted. 'He knew that Shockley had the intellectual versatility and the brilliance to do it,' said physicist Conyers Herring, who worked at ASWORG.[35]

Shockley stood very little probability of getting drafted, he told May in summer 1940. Being 30 years old and a father put him well down on the list. Moreover, he assured her, no PhD at Bell Labs ever was likely to get called up or even be allowed to enlist. When the forms came in from Selective Service, the recipient sent them up to the labs' administrative office and the labs told the draft board the man was much too valuable to the war effort at the bench. It worked every time.[45]

According to folklore, most physicists do their best work before they reach 35. Shockley was 32. Although he worked at a premier industrial laboratory in an exhilarating field that showed great promise and approaching what could be the apex of his creative life, Shockley knew he, like everyone else, was about to be side-tracked. Shockley could not question the necessity of the diversion; he knew his duty and supported the war effort. Although he voted for Wendell Wilkie against Franklin Roosevelt in 1940, Shockley considered himself half-British. He spent considerable time circulating a petition around the labs, demanding a speed-up of America's rearming.[29] Nonetheless, his life and work faced disruption just as he approached the peak of his intellectual fertility.

Bell Labs put him on leave, and Morse immediately assigned him to the problem of dud depth charges. On the rare occasions that US Navy planes spotted U-boats on the surface, they dropped depth charges on the fleeing subs. The charges exploded but seemed to have no effect: the boats managed to submerge undamaged. That struck the Navy as peculiar since the same devices were highly effective when they were dropped off the side of destroyers.

It took Shockley only a few days to figure out the problem. The Navy set the charges to explode at a depth of 75 feet when they were dropped from destroyers. Since that seemed to work fine, the charges put on planes had the same setting. When the destroyers used the charges, the subs were submerged; when the planes deployed them, the subs were on the surface. The result: the charges dropped from the air exploded too deep to cause damage. Shockley's simple suggestion: set the charge on the airborne devices for 35 feet, not 75. Within two months, Shockley's idea had increased the number of U-boat hits by a factor of five.[37]

Now there was the problem of finding the submarines in the first place. 'Our reasoning went something like this,' Morse wrote about the general problem facing ASWORG. 'If the submarine is dangerous because it is hard to find, then the process of finding the submarine is an important part of the counteraction.' To help solve that problem, however, they needed real information about the antisubmarine efforts, and they quickly found the reports the Navy provided were anything but. Sailors and officers in the field filled out the forms willy-nilly. The sailors put in as little time as they could filling out the forms, sometimes making up the entries. This was particularly true in reports about submarine hunting. The Navy flier who famously remarked 'sighted sub, sank same' lied. The same sub sank a ship the next day.

Most of the time, the German U-boats sailed on the surface, sucking in air to charge their batteries and to communicate with their home base, usually in bursts of coded radio signals. To help the Navy find those subs, ASWORG had to know about the crews, the planes, the tactics, the equipment of the antisubmarine forces. They could not learn this from the room-full of sloppy Navy reports; they had to be out in the field, just as Blackett's minions roamed the RAF and Royal Navy bases in Britain. By the end of summer 1942, half a dozen ASWORG scientists, civilians all, were stationed in Atlantic and Gulf coast bases. The ASWORG scientists at the bases talked to the men, inspected the equipment, and even flew missions. A few were wounded, only one seriously. Several won medals.

The chain of command to ASWORG went from the President of the United States to the Office of Emergency Management, to Vannevar Bush's Office of Scientific Research and Development and then to NDRC, which was being run by Harvard's James Conant. NDRC had three units under it, one on ordnance (presided over by Caltech's Charles Tolman), another on instruments (MIT's Karl Compton), and communications and transport (Bell Labs's Frank Jewett). Bush had enlisted American science's 'old boy network' to fight the war. NDRC then had a contract with Columbia University, one of several side contracts for ASWORG, and Columbia wrote the checks for Shockley's organization.[38]

ASWORG comprised scientists and actuaries. 'The probability of finding a submarine is proportional to the area swept out over a given amount of time, and the probability of the submarine resurfacing is a function of time,' said Conyers Herring, another member of the group. 'You just use mathematical probabilities and add them up. They are relatively trivial things but things people wouldn't think of using unless they were used to using probabilities. For that reason, a very large number of people who were recruited for the research group were not scientists but actuaries.'[35]

Shockley's unit virtually exploded with ideas, some of them so simple and clever that the Navy's best weapon early in the war against Germany's submarine threat was a couple of dozen men with pencils and paper. Modern warfare, it turned out, could be turned into charts.

Problem: what is the most effective way to search by air for submarines?

The First Bomber Command, operating out of the East Coast, wanted to know whether planes with radar really did significantly better than planes relying on visual sighting; whether sightings of U-boats depended on how far from shore the planes flew, and whether faster planes were more efficient in the hunt than slower ones. The radar question was interesting because the ASWORG scientists in the field discovered that antisubmarine crews, fearing the Germans had radar detectors, frequently turned off their radar so as not to be discovered as they closed in on surfaced subs.[37]

The scientists broke down the average hours of flying per sighting by whether the planes had radar and whether they patrolled within 60 miles of the coast: the higher the number, the less efficient the search. Then they worked out the probability their data were correct (they were) and told the Navy that using radar tripled the chances of its crews finding a sub, and the subs – if they had radar detectors – weren't turning them on. Add to that the obvious advantages of radar at night and in bad weather. The Navy ordered radar for all its planes, and ordered its fliers to keep the radar on. Flying beyond 60 miles clearly was more efficient than staying near the coast, ASWORG also found. Surprisingly, the ASWORG scientists also proved that slow-moving planes were more efficient than faster aircraft, probably because pilots had to pay more attention to flying the speedier planes and less to looking for subs.[39]

One survey that Shockley's group performed that may have saved hundreds of lives was a statistical answer to another serious question: Did the German bombers targeting the Atlantic convoys have radar? The bombers were responsible for 30% of the convoy sinkings. If the German bombers did not have radar, then the convoys could hide from them under clouds or in fog and were relatively safe at night. If the planes used radar, there was no place in the wide ocean to hide.

The Allies occasionally got their hands on a bomber, and none of them had radar, but they could not extrapolate from those few captures that the rest of the German squadrons did not. Perhaps the reason these planes were captured was that they were the ones that didn't have radar.

First, Shockley's group used data obtained from British listening devices planted all along the Channel coast to detect incoming aircraft. They could hear planes taking off and landing on the German-held European mainland. From that data, they deduced how many hours the bombers spent in the area where convoys sailed.

Next they obtained meteorological data on the average visibility in the convoy area. This helped them to figure out just how visible the convoys were from planes above the ships. Then they calculated, from the hours searched and the weather during that time, how many convoys the Germans would be expected to spot from the air relying only on the eyesight of their crewmen. Using Navy data on actual convoy attacks, ASWORG discovered that the number of attacks almost exactly matched the number expected from visual surveillance. If the Germans used radar, the number of sightings would have been dramatically higher. Conclusion: the German bombers did not have radar and convoys might find places to hide.[40]

Shockley and Morse continued building their organization. Headquarters moved to Washington and room 4807 on the third floor rear of the 'Navy Main.' One room in the complex remained off-limits to the scientists and even to the petty officers that acted as administrative staff. Here the Navy kept its most secret documents; the room that grew in mystery as the months passed.

The work put Shockley on an almost perpetual traveling schedule, spending long hours on trains, going from Madison to New York to Washington or Boston, then flying to points south. He had a top priority pass, signed by the secretary of war, to get on any commercial plane he needed. His orders always were specific since he carried classified information. 'You are directed to carry two hundred pounds of excess baggage,' one order read. 'One hundred and twenty pounds are to be carried in the baggage compartment and eighty pounds are to be carried in the passenger department in your personal possession as these are secret and confidential property of the Unites States government.'[41]

At 3 p.m. on 23 August 1942 Shockley's first son, William Alden, was born while Shockley was away. 'He's a honey!' Jean reported to May. 'He has lovely flat ears, like Bill's – not like mine and Alison's, and pretty hands and a pretty shaped mouth.' Shockley took the overnight train down to see the new baby. 'He is just as thrilled as I am at having a son this time,' Jean wrote. He then drove up to Alison's summer camp in Sussex County to tell her about her brother.[42] Then, he left on another trip.

May sent bouquets every day for four days until Jean wired her to stop.

Shockley's schedule meant he saw very little of Billy. He took a permanent room at the University Club in Washington, so he had regular access to a pool and gym, and spent most of the summer working on radar problems, particularly the reflectivity of various metals.

Figure 11 Alison and her brother, Billy, in New Jersey.

In the first week of December, he and Morse left on a long trip to England to visit Blackett. Morse was curious to see how Blackett had made the transition from noted physicist to war scientist and administrator. Morse's excuse was the dispatch of a US Army Air Force bomber squadron to England to help the British Coastal Command fight the U-boats. The British had been blasting the subs' dock pens in northern Germany to no apparent effect. The subs kept coming. What appeared to work were nighttime sweeps by radar-bearing bombers that caught the subs as they raced for the open sea. That required resources the British did not have, so they asked for American assistance.

Shockley and Morse wanted to see how Blackett's scientists helped the military react to changing situations. They took the safest and fastest route to London: a Pan American Clipper from New York to neutral Lisbon, thereby avoiding the danger of crossing the ocean further north. The flying boat refueled in Bermuda and the Azores, where they waited until the water calmed down enough for the Clipper to take off. They landed in Lisbon, but only after circling long enough for the fishing boats to get out of their way. Shockley and Morse flew to London the next night, taking the long, over-water route with lights out and radios silenced to keep them invisible to patrolling German fighters.

London was blacked out. Cars navigated the twisted streets at night with headlights dimmer than fireflies, Morse wrote. Walking the streets was an adventure. 'Cheerful, iron-nerved girls in uniform' drove them from meeting to meeting.[37] The Germans bombed.

Shockley wired May: 'WELL AND COMFORTABLE STAYING MONTH LONGER VICTORIA NOW BUSINESS DISTRICT HAVE DUAL NATIONALITY HERE.'[43]

The man they went to see developed the use of statistics as a weapon of modern warfare. Blackett called it 'operational research,' which in the US now is called 'operations research.' In fact, operations research preceded Blackett by 100 years, the invention of the wildly eccentric British father of computing, Charles Babbage, in his textbook *On the Economy of Manufactures*, in 1832. With stopwatch in hand, Babbage measured the efficiency of manufacturing – he was the first time-study man – detailing how things got made, everything from books to needles. He put down the numbers and then organized the data to draw conclusions not obvious even to the people doing the work.

Patrick Maynard Stuart Blackett was born 65 years later in Chelsea, from a family of clerical and maritime roots. He served as midshipman during the First World War, participating in the Battle of the Falklands, where his ship was heavily damaged. After the war the Navy sent him to Cambridge for further training, but Blackett stayed on as a civilian, eventually doing research with Ernest Rutherford in the heady birth of atomic physics. Rutherford, then studying the effects of firing alpha particles into the nuclei of atoms to see what popped out, asked Blackett for help. In the 1920s and 30s, Blackett perfected the design of the Wilson Cloud Chamber, which won him the 1948 Nobel Prize in Physics. Despite their success, Blackett and Rutherford fought often. Blackett left for Birkbeck College at the University of London to study cosmic rays; then in 1937 he moved to the University of Manchester to work on the Earth's magnetic field.

Blackett loved committees. He loved heading them or just being on them. One he joined, the Tizard Committee, began in 1934 studying ways in which scientists could help Britain should she ever be dragged into another war. Blackett quickly got into a brawl with a member, F. A. Lindemann. Blackett wanted the committee to study radar; Lindemann wanted to use infrared detection. That drove Blackett away, but his early work led to Britain's effective deployment of radar against the Germans, which helped win the Battle of Britain.

When the Second World War broke out, Blackett went to the Royal Aircraft Establishment at Farnborough and then to the anti-aircraft headquarters at the Coastal Command, and finally to the Admiralty where he could form his own committee. The problems scientists faced in the war, Blackett wrote in a memorandum marked MOST SECRET, were more like biology than physics because 'a limited amount of numerical data is ascertainable about phenomenon of great complexity.' Past operations were studied, analyzed, and then 'are used to make predictions about future operations. This procedure ensures that the maximum possible use is made of all past experiences.'[44]

For example, using statistics from the previous war, the committee-phile Blackett quickly developed Blackett's Law: 'The number of ships [in a convoy] hit is inversely proportional to the size of the convoy.'

One ASWORG member, botanist Kenneth Thimann, discovered the same principle. Nature protects schooling fish that way: Predators attack the school from the outside so fish on the perimeter provide the first targets, protecting the center. The larger the school, the lesser the odds that any one fish will get eaten and the greater the proportion that will survive. Similarly, by increasing the size of convoys, precious goods got through and fewer mariners died.

Blackett happily escorted his visitors around England, including a trip to the American bombing command, which already had a unit similar to ASWORG. That group, in the UK, working for the Army instead of the Navy, had a different operating philosophy, in part because it was run by a lawyer, Harvard's W. Barton Leach, instead of a scientist. Leach let the Army control his scientists in the field, working as an adjunct to the commanders. He only began projects when the commanders approved. One of Leach's men, Robert Robertson, found himself fruitlessly engaged in an internal debate over British and American plans to bomb German cities. His research suggested the bombers would hit far more homes than military targets and the Germans were likely to react the same way Londoners did to the Blitz, with growing, angry resistance. His efforts were for naught; his commander, another lawyer, wouldn't allow Robertson contact with his superiors.

Morse and – at first – Shockley believed their system better; the Navy gave them more freedom to see what they needed and they retained control of their personnel. Further, they had access to the Navy brass and an objector such as Robertson would not be isolated within ASWORG. Shockley later changed his mind.

Shockley gathered as much information as he could about the crews and their performance. He discovered that the 'lifetime' of a crew lasted for only one chance at a German sub. Statistically, before they got another shot at a sub, someone on the crew would have been transferred, killed, or wounded. Crews had very little time to learn how to work together as a team and this adversely affected their efficiency. There was no such thing as an experienced antisubmarine crew.

His data led to two conclusions. First, ASWORG needed to provide the missing wisdom. By studying every crew and all the reports of U-boats, they would provide some of the knowledge the crews lacked by not being together long enough to get it themselves. Second, training was crucial. Shockley eventually spent a large chunk of his wartime efforts improving the training of flight crews, first the antisubmarine crews of the Navy, later the radar bombing crews of the Air Force. It was his greatest contribution to the war effort.

Just after Christmas, Morse returned to Washington aboard the *Queen Elizabeth*. Shockley remained in Britain for another month.[37] Jean, and more certainly the children, missed him at Christmas time. 'I think Christmas this year would have been perfect if Bill had been home,' Jean wrote sadly to May. 'Our tree was the prettiest in years because we decided to get a "Billy-sized" tree that would stand on the card table. Consequently, the ornaments and the four strings of lights twinkled in more concentrated splendor, and the bright little tree (about five-feet-tall, not microscopic, at that!) just outdoes itself in shining.'[45]

Shockley kept finding things to keep him busy, being in great demand by the British. 'Jobs sprout along Bill's path like mushrooms,' Jean said to May, passing on one of his letters, '...and he is very busy. I shall include weed-killer in his equipment, the next time he goes on a trip.'[46]

The trip greatly stimulated his creativity, and he no sooner arrived back in Washington than memos exploded from ASWORG, some of them composed on the way back, typed on British stationery. The spring and early summer of 1942 constitute one of the most productive periods of Shockley's life. He concentrated his research attention on the problem of efficiency. Armies are by nature inefficient; if you have a problem, you simply pour men and equipment at it and overwhelm the obstacles, providing you have the resources. Using Blackett as a guide, Shockley thought the war could be better fought if the military could measure just how efficient certain operations were, and then find ways to increase that efficiency. In many ways, the first problem was the hardest.

In arguing for new statistically based efficiency tests Shockley used industrial analogies, describing how the Bell system developed and trained field personnel and eventually deployed new switching stations. Then he compared that with how the military ran its training for new technologies. Soldiers in the field ran months, perhaps even a year, behind war technology. When the technology got into the field, the people most needing to use it didn't understand how it worked. 'The writer believes that a large portion of this delay is unnecessary and could be eliminated by a proper procedure for the development of measures of effectiveness,' he wrote. He designed a program to do just that.

His unit reported to Edward Bowles, an MIT professor, on plans to engage in a proposed tactic known as 'exhaustion hunt method' for sinking U-boats. In this strategy, once a plane discovers a submarine, planes would swarm to the area to constantly harass it, making it impossible for the U-boat to escape or surface to recharge its batteries. An exhausted, suffocating U-boat crew then had the choice of death or surrender. Shockley concluded that the effectiveness depended on where the Navy tried the tactic, but if used properly, exhaustion techniques promised a great increase in the number of sub hits.[58]

This study led the ASWORG hierarchy to uncover a deep secret of the war. Shockley's group wanted to know the reliability of the data on submarine location. The Navy spotted the subs when they surfaced and blasted off quick coded messages to Germany. Radio direction-finding (RDF) devices triangulated on the broadcasts and officers in the secret room at ASWORG dispatched planes to the area.

ASWORG's Jay Steinhardt found the location given to the planes was usually exactly right – much too right. The technology could not possibly be that accurate. When Morse reported Steinhardt's discovery to the Navy, he received a blank stare. The next day he was let in on the secret: the British had broken the German code. They not only located the radio transmissions, but could listen in as the sub commanders sent the sub's exact location home.[37]

Throughout the year, as Shockley absorbed himself in his work, tensions grew at home. He had seen little Billy only a few times. Jean constantly lamented her inability to even correspond with her husband during the war. 'Bill's comings and goings are so uncertain (ditto the hotel he may be staying at) that I should say the chances of a letter addressed simply to his room number finding him would be quite small,' Jean told May. 'Secrecy pertains mainly to his work, so far as I can see – not to his whereabouts or identity, as long as he is in the vicinity of the eastern seaboard.'[47] That changed, and even his whereabouts became a military secret.

Shockley took one break back with Jean during this blizzard of memos, perhaps to ease the disharmony. He and she went off to the St George Hotel in Brooklyn for a rare vacation. They had hoped to go to a Georgia Sea island, but by the time Shockley discovered he could take time off, the hotel was booked, so the salt-water pool and sun lamps at the St George had to suffice. The children were left with a sitter. They spent the evenings at the movies and nightclubs of Manhattan. Jean was highly amused watching the dating couples and the games they played with each other, 'a short course in How Did She Do It, with opportunities for the close study and observation of techniques.' Shockley had been with the military long enough to read uniforms and insignias, and as men passed by their nightclub table, he could identify their units for Jean. The city throbbed with servicemen, mostly, it seemed, from the US Navy. They went to a ballet at the Metropolitan Opera House on their last evening.[48] Tensions, for a brief time at least, eased.

Two months later, Shockley spent a few days in San Francisco and saw May. Jean knew he was in California only because she got a telegram from him saying so. 'I'm trying to really demand more of his time on the rare occasions when he has a day at home, but when he is away I don't keep track of him at all,' she wrote. When he did make it, she tried to get him not to bring his work, a plan that did not always work. He stared off into space, totally absorbed. If she asked what he was thinking, all she got back was a sheepish grin. He could not share his work with her; much of it was strictly secret. Every few minutes the telephone rang. Jean told him next time she had to pick up the phone she would identify the number as 'Lonnegan's Bar & Grill.'

She saved her ration stamps so they could have meat on his days home.

Part of his concentration was on his new role as organization leader, which Jean describe as the 'executive bends.' She defined it as when you get into the rarified atmosphere of administration characterized 'by constant headaches and shooting pains in the neck.' Shockley, she wrote, was reading a book on management.[49]

The executive part of the job also worried Shockley, who despite his excessive self-confidence, had never run an organization or led a team before. He was never burdened with self-awareness. Largely deprived of peers during his childhood, he lacked experience with other people,

Figure 12 Shockley and unnamed officers during the Second World War.

particularly people unlike himself, especially those less bright, which was almost everyone. Worse, he did not realize the extent of this weakness and would pay dearly for that ignorance.

He spent considerable time recruiting for ASWORG and helping the Navy assign the right people to antisubmarine warfare. Finding the right personnel particularly depressed him: so few measured up to the standards he set, officers and enlisted men alike.

His job was particularly delicate because the scientists in his group and the military for whom they worked represented two entirely different cultures. Clashes were inevitable. Each newcomer received a pamphlet describing the potential conflict written in a style that is pure Shockley. He required every member of the team to reread it every month. 'Our job is to help win the war, not to run it ourselves,' he warned recruits. 'We begin to be useful when we can combine with our scientific training a practical background gained from contact with operating personnel. This practical background can only be obtained when the operating personnel trust us and like us.'

The main problem was convincing the Navy that 'scientific work done without interchange of ideas between workers in the field is a contradiction in terms.' The kind of interchange needed flew directly into the face of the military culture and Shockley spent the war trying to find ways of making peace between the warring cultures.[50]

ASWORG quickly expanded to 48 men, spread over 14 sites in England, Cuba, Iceland, Morocco, and Brazil, with nine, including Shockley, based in Washington. Shockley and Morse essentially served as the administration.[37]

The University Club in Washington became his home, the pool allowing him to get in his laps before breakfast. His arrangement with Jean was to stay in Washington from Wednesday through Saturday night. At six on Saturday, he took the train to Madison, arriving home about midnight. He stayed home Monday and Tuesday and then returned to Washington on Tuesday night. Only a one-week trip to Cuba, Miami and Key West on business in late June disrupted the schedule. He gave up any attempts at gardening or doing repairs around the house and lamented the death of his lawn. Jean ran the family's Victory Garden; Shockley was too distracted. After he left one day, Alison told Jean little Billy just didn't know his father was a member of the family, 'he comes home so scarcely yet.' Jean described him as a comet periodically visiting the solar system.[51]

Shockley was an unhappy comet. On 6 November 1943 he came home for one of his three-day stays. The weather was unusually warm and sunny for early November. He and Jean, as was their habit, took long walks through the woods beyond the town. Two days after he arrived, Shockley sat and wrote her a note.

Dear Jean:

I am sorry that I feel I can no longer go on. Most of my life I have felt that the world was not a pleasant place and that people were not a very admirable form of life. I find that I am particularly dissatisfied with myself and that most of my actions are the consequence of motives of which I am ashamed. Most people do not feel this way I am sure. Consequently, I must regard myself as less well suited than most to carry on with life and to develop the proper attitudes in our children. I see no reason to believe other than that I shall continually become worse in these regards as time passes.

I hope you have better luck in the future.

Bill[52]

He took out his revolver, put a bullet in one of the six chambers, put the gun to his head and pressed the trigger. Nothing happened. He put the gun away and wrote a second note.

Dear Jean:

There was just one chance in six that the loaded chamber would be under the firing pin. There was some chance of a misfire even then. I am sorry that I was not sufficiently ingenious or painstaking and find a more practical and suitable means of solving our problems.

He secreted the suicide note and its companion in an envelope in a safe in his home along with other personal material found after his death.

His suicide attempt was serious, even with the end determined by a game of Russian roulette. Suicide experts say that Shockley's attempt was not unusual. Religious people let God choose life or death for them. Rationalists, such as Shockley, let chance decide. Either way, the person has decided that ending his or her life was appropriate or desirable. Even his note is typical of 'completers,' people who are serious about taking their own lives. Real suicide notes are short and to the point. People who are not serious generally write long, rambling farewells.

So far as is known, he never attempted suicide again, and this too is not unusual. Psychologists say that many people, after surviving one suicide attempt, never try it again: God or a random universe has given them an extension on life.

What provoked this attempt is unknowable – what was said, what happened. His unhappiness was reflected in his surviving letters, which while they never showed emotion, became shorter, more rare and even less personal. But the letters did not bode such extreme depression, and he told no one about the attempt to kill himself. Since the letters were not found in a posted envelope, it can't be certain Jean ever knew about them or read them.

Unlike Jean's pacific descriptions to May of his weekends at home, the two fought often. Having now two children, including a son he rarely saw or barely knew, made the situation worse. Alison remembered tensions, but not angry words.

Her children describe Jean as plain, unexceptional and physically unaffectionate to Shockley or to them. Shockley, like his mother, was quiet, self-contained and physically unaffectionate, to Jean or to them. He spent most of his waking hours at a job that required his total attention, playing on a stage of high, historic drama where human lives in the millions – not to mention civilization – were at stake. He inhabited a world of night-time flights in hunted planes painted black to avoid predators, of men who went to work with an excellent chance they would die gory deaths before they returned. Then he came home, where changing a diaper was an event, to a companion whose conversation, unavoidably, was usually limited to ration stamps and whether they could afford to buy their daughter a new coat. Many men appreciated that kind of respite, grateful for the mundane, calming reminder of how life ought to be. Shockley could not.

Jean's description of Bell Labs being an ivory tower was more apt than she knew. In many ways, places such as Bell were far more isolated from the real world than were universities. Schools have a steady stream of students going through, refreshing the air and trailing in with them whiffs of reality. Not so private laboratories, where, except for the hired menial help, the only contact the scientists had with other humans was with bright, highly educated, literate, focused people just like themselves.

Shockley was now thrown into the military, which during the Second World War represented as good a cross-section of male America as ever existed. Some officers he met were bright – none as smart as he, to be sure – and dedicated. Many were not, and Shockley could not like them very much. He had no patience for mediocrity.

He began spending more time with the Army air wing because of his growing dissatisfaction with the Navy's policy of withholding secret information that he believed he needed. By then, the Battle of the Atlantic had been won and he may have become bored with the Navy. He flew on training missions with Air Force crews, and by January 1944 he was working full-time for the Air Force and Bowles' high-level consulting group.

Shockley's first task for the Air Force was devising general training procedures for bomber crews equipped with the new radar out of RadLab, called Plan-Position Indicator scopes (PPI), which displayed the echoes as on a map. Shockley flew over North Carolina and took the pictures for his instruction manual, later distributed to trainees in a dark maroon cardboard cover. Each picture represented what the operator might see on the screen with a description of true and false images.[53]

On 13 September 1944, Shockley left on his longest trip of the war.

For any other man, traveling around the world during a world war as a high-ranking civilian with considerable clout would have been one of life's greatest adventures. If Shockley appreciated the trip that way, it does not show in either his letters home or in the little black, government-issue daybooks he carried. The excitement of such an exploit seemed to elude him. Life had begun to break down into a series of intellectual exercises and challenges and he was distracted from everything else.

Incidentally, the first book notes that the US had broken the Japanese code. This entry is surprising for two reasons. One, the breaking of the code was one of the best-kept, most strategically crucial secrets of the war, for it meant the US could listen in to Japanese military communications. That Shockley was let in on the secret suggests how much he was trusted and how important the military thought his work. Two, that he wrote it down in his daybooks, which could have been captured or lost, seriously violated security regulations and showed either foolhardiness or a deep lack of respect for the regulations. Had Shockley's books been captured by the Axis, the course of the war might well have been altered, because neither the Japanese nor the Germans knew their codes had been broken.

First stop: England. 'Bull sessions on targets. Folders for target study, some records... space for operators to initial,' he wrote in his daybook. He set up training runs and tests of two models of radar. He had one crew fly three days over German targets to test out his procedures.

Within a few weeks, he was on his way to India. 'I had quite a trip over here,' he wrote back. Most of the flying was done at night to avoid detection and, as he noted, the planes weren't designed for sightseeing. 'I have a pretty incomplete picture of what the country is like.'[54]

'I was struck by the monotony of the scenery,' he said years later. 'As far as the eye could reach from the low-flying transport airplane, I was surrounded by rice paddies which stretched out into a continuous plain, much like an ocean of grass. Occasionally, in this ocean, small islands in the form of clumps of trees arose. The trees represented villages of mud houses.' On one plane trip, he saw it as a giant puzzle and wondered if rearranging the geometry of the patterns would increase rice production. He concluded the difference would be negligible – 1 or 2%.

He saw people in depressing numbers, especially in Calcutta, and distressing poverty. Years later, he still couldn't get the lamentable scene out of his mind.[55] Shockley also saw elephants, monkeys, jackals, and scores of Buddhas, and visited restored temples – those recorded with slight attention and only passing interest. Others might see such exploration as fun. Shockley could not.

He traveled back and forth to the air base for weeks at a time. The commuting led to what he called the Karagpur Effect: 'action does not equate with progress.' He was traveling hundreds of miles a week and getting very little done.[56]

On 28 November he headed for the Pacific Theater of Operation. First, a stop in sweltering Brisbane, Australia, after two weeks of travel.[57] From there Shockley flew to recently liberated Saipan in the Pacific to study the training and efficiency of the new B-29 Superfortress bombers, now blasting the Japanese mainland.

Wartime censorship limited what he could tell his family, and he made every effort not to frighten Jean or May. In one letter, dated the day after Christmas 1944, he omitted the fact – but noted in his daybook – that while he was playing chess, a Japanese bomber flew over, picked its way through the anti-aircraft barrage and bombed the base, Shockley's only combat experience. He spent part of Christmas afternoon in a trench.[58,59]

From Saipan, he turned back to Ceylon. There, Shockley helped set up the radar charts for the bombing raids on Osaka and Nagoya. He also fired a blistering note back to Washington when he found that some equipment being sent to the Pacific had defects that should have been spotted in the factory.[59]

When he returned in February 1945, Bowles, acting apparently on orders from General H. H. 'Hap' Arnold, chief of the air staff, asked Shockley to take on some larger issues. Shockley's official title became Expert Advisor to the Secretary of War. He had the power – and used it – to order Army generals to attend meetings to decide future research.[60] He also returned to Bell Labs part-time, organizing the solid state physics group at Kelly's request.

First, Shockley produced a secret document on ways in which quantitative techniques, such as those he and Blackett exploited in the European theater, could help finish the war in the Pacific. He pointed to several tactical mistakes Blackett found in European operations, including the bombing of German cities, which proved to have far less effect on the German war effort than expected, just as Robertson had predicted. Shockley wanted to make sure similar mistakes weren't made against Japan.

Blackett found that all the bombs dropped on Germany up to July of 1944 destroyed only about 10% of its wartime industrial capacity, and that two months after each raid, the targets were back to normal production. The Germans' factories totaled 110 square miles of space. Bombing destroyed only 7.2 square miles of it even after years of aerial bombardment. Attacks on specific industries, such as the factories building the single-engine fighters and the synthetic facilities factories, did considerable damage, but even those raids had less total effect than first believed.

Overall, Shockley said, the air war on Germany was 'profitable,' but less so than appeared. For every ton of bombs dropped on Germany, Shockley reported, the Germans lost between 52 and 122 man months of effort. Each ton of bomb cost the British 18 man months. The result was profitable by a factor of three to six times for the British Lancaster bomber.

Shockley then applied his cost–benefit analysis of aerial bombing to the Pacific. The US lost 6% of its new, huge B-29s every time it sent planes over the Japanese islands. On each raid, one plane carried three tons of bombs, which equaled 120 man months per ton. Casualties came to one crewman per five tons of bombs, which cost the Japanese 500 man months of labor, 'that is, we lose one man for a gain of about forty man months of Jap labor,' he wrote.

'These figures are very striking and on face value, throw doubt on the soundness of the B-29 program,' he wrote. It might be possible to make the B-29 raids more profitable, but Shockley urged a serious study of the options. Perhaps concentrating on bombing Japanese shipping would be more profitable than continuing attacks on Japanese cities. Besides, the Air Force was running out of 'fresh targets' in the cities.

On 21 July 1945 he proposed to Bowles a major study on the casualties that might result if the Allies invaded the Japanese islands – Japanese casualties as well as Allied. From the tone of the report, it was not clear how much Shockley knew about the atomic bomb program. Virtually every physicist in America – including Shockley – knew about the Manhattan Project even if only by a process of elimination: almost every nuclear physicist in the country mysteriously disappeared around 1942 to work in a remote section of New Mexico or at Fermi's lab in Chicago on a project they couldn't discuss. Shockley knew that Seitz, for instance, worked on the project; they had dinners together and carefully avoided talking about work. Shockley had no reason to know the decision to use the bomb had been made.

Shockley said that a major study on the effects of an invasion had never been done, but after looking at what happened elsewhere in the war, he felt that the carnage would be appalling. The few studies completed on past operations against the Japanese in the Pacific indicated that for every ten Japanese soldiers killed an American died. The ratio was oddly consistent throughout the war. 'We shall probably have to kill at least five to 10 million Japanese. This might cost us between 1.7 and 4 million casualties including 400,000 to 800,000 killed,' Shockley concluded.[62]*

If Shockley's July memo was circulated in the Pentagon – and historians have not found any other copies except in Bowles' archives – it probably only confirmed a decision already made to drop the bomb on Japan. His research on the ineffectiveness of the B-29 raids in February was a different matter. The current bombing campaign was less effective than the air force had hoped, Shockley found, and, as he suggested, options had to be considered. Whether he actually meant the atomic bomb will never be known. Because Shockley had a clear line directly to 'Hap' Arnold, commander of the Air Force, his report was unlikely to be ignored.

Two weeks after he dispatched his memo, the atomic bomb was dropped on Hiroshima and three days after that on Nagasaki.

Shockley's reform of training with radar bombing crews made it possible for the US Air Force to begin night bombing of Tokyo, Nagoya, Osaka, and Kobe in March 1945. The planes flew at 7,000 feet, dropping mostly incendiary bombs. In the first attack 1,667 tons of bombs fell on Tokyo alone.[61]

In August 1945, General Curtis LeMay looked at pictures of a night air raid on the Maruzen oil refinery at Shimotsu and found the Air Force crews destroyed 95% of the refinery. 'This performance is the most successful radar bombing of this command to date,' LeMay wired the commanding general of the 38th Flying Training Wing in Arizona. At the end of 85 hours of Shockley's training regimen, the crew could hit any target within 1,700 feet – at night, at high speed and low altitude. When

---

* Stanford historian Barton Bernstein said that a study of potential casualties had been done a month before, without Shockley's knowledge apparently, and it came up with much fewer casualties on both sides and that Shockley's admitted guess was greater by a factor of ten.

they missed at all, it was because of inaccurate location on the aiming point on the radar scope.[64] 'Hap' Arnold credited Shockley with directly influencing the winning of an early victory over Japan.[65]

Impressed, Arnold enlisted Shockley informally to his staff and asked him to ghost-write a chapter under Arnold's byline for a small book published by the Federation of Atomic (now American) Scientists called *One World or None*. The literary company was heady for an Army general: other contributors included Einstein, Oppenheimer, Szilard, Wigner, Condon, Walter Lippmann and Fred Seitz, who had moved to the Carnegie Institute of Technology in Pittsburgh.

In the chapter titled 'Air Force in the Atomic Age,' Shockley wrote that massive air attacks, such as the ones that obliterated most of the Axis cities, made mass destruction cheap and easy even before the atomic bomb. This rendered 'the existence of civilization subject to the good will and the good sense of the men who control the employment of air power. The greatest need facing the world today,' Shockley wrote for Arnold, 'is for international control of the human forces that make for war.'

Shockley, at Arnold's request, had begun doing research on the cost–benefit of the atomic bomb, and he incorporated some of this work in Arnold's essay. The Germans proved over Coventry that a city could be destroyed by high-explosive bombardment. The atomic bomb, however, made such an act dramatically cheaper. Each bomb cost $1 million; delivering such a bomb to one city cost $240,000. The bomb that hit Hiroshima destroyed 4.1 square miles and the Nagasaki bomb destroyed 1.4 square miles – an average of 2.8 square miles per bomb or less than a half-million dollars per square mile obliterated – six times cheaper than using conventional weapons. That price, Shockley noted, was bound to get even lower as bombs got bigger.

Shockley then pointed out that the monetary cost to Japan ran to at least $160 million per square mile destroyed, which made atomic bombing 'profitable by a factor of fifty; that is, the cost to Japan was fifty times the cost to us.' He projected that in the future 'every dollar spent in an air offensive is expected to do more than $300 worth of damage to the enemy.'

With one exception – a reference to the fact the Air Force could have theoretically wiped out 21 million Japanese in 68 cities in one afternoon using nuclear weapons – the Shockley–Arnold essay did not mention the human cost of the bomb, the death and horror.

Much of the science of operations research in the US derives from Shockley's work for the military in the 1940s. He made it respectable and proved it could have serious practical utility.

Continuous letters of praise for Shockley's war efforts poured into Bell Labs, along with a request he remain as a part-time advisor to the Pentagon. The military wanted him to continue his operations research, especially into the atomic bomb and rockets. Bell Labs agreed and Shockley worked for the Pentagon as a consultant for much his life.

Shockley's war efforts represent in some ways his finest moments. It was the last time his professional achievements were clear and unambiguous. Nothing else would ever be as completely satisfying. And his failures would be astonishing.

On 17 October 1946 the 12-year-old Alison and her father took the train to Washington, where Shockley was awarded the National Medal of Merit, the highest civilian medal honor.

Shockley wrote to May that five major generals and a few brigadier generals showed up for the ceremony. He promised to send her a picture of the medal. Oh, yes, he said in one matter-of-fact paragraph between descriptions of the ceremony: Jean lost the baby. Singular.

In fact, Jean had been pregnant with twins.

While Shockley and Alison were away, she had gone into premature labor and was rushed to a Summit hospital.[66] The doctor told her she had miscarried one of the fetuses and the other was dead. He had to extract the dead one immediately, he said, and reached between her legs while she screamed.[67]

CHAPTER 5
'I think we better call Shockley'

Thomas Alva Edison was no scientist. Indeed, he had little respect for scientists. He was a practical man and most researchers spent their lives in impractical pursuits. Edison did only one real piece of science in his life, and that unintentionally.

In 1883, shortly after he invented the light bulb, Edison began tinkering with the bulbs to find ways of improving their efficiency. His bulbs burned out too quickly. The white-hot filament that provided the glow charred and grew progressively thinner, finally snapping with a flash.

His filament was a horseshoe of carbon wire sealed in a modest vacuum in a glass bulb. Edison noted that as the filament burned and turned black, a thin white line appeared on the glass opposite one leg of the filament. The line seemed to have been produced by something emitted from the far leg of the filament, something that was being blocked by the near leg. Puzzled, Edison tried sealing a metal plate upright parallel to the filament. He saw a most perplexing thing: When the plate was connected to a positive charge, a current flowed from the negatively charged filament to the new plate without the two touching. When it was connected to the negative charge, nothing happened.

He reported what he found, took out a patent on the tube in 1883, and admitted he had no idea what was going on or what good it would do anyone. Edison surmised that since the effect only occurred with a negatively charged filament and a positive plate, whatever was producing the current also was negative – opposites attracting and all that. Since science had not yet discovered the electron, neither he nor anyone else could explain the phenomenon.[68] A British engineer replicated the experiment a few years later and called what he saw the Edison Effect.

Then in 1897, the British scientist Sir Joseph John Thomson found the electron, a negatively charged subatomic particle. Five years after that, Owen Willans Richardson put it together: He found that a metal filament, when heated in a vacuum, evaporates electrons. That accounted for the flow to the positively charged plate – the Edison Effect.

Four years later, John Ambrose Fleming proved Edison wrong: the effect did have a practical use. Fleming inserted a metal plate around the filament. Then he attached a source of alternating current, current that changes direction many times a second. When the current moved one way, electrons sailed off the filament to the plate. When the current reversed, nothing happened. In other words, Fleming's tubes made electricity flow in one direction only, from the filament to the plate. He was putting alternating current into the system and getting direct current out. The tube acted as a valve to the current.

The process of converting alternating voltage to direct current is called rectification, so Fleming's valve was a *rectifier*. It was the world's first electronic device.

Because it had two elements, the filament and the plate, it also was called a *diode*. The tube also could receive wireless (radio) oscillations and pass them in one direction. Since Fleming worked for the Marconi Co., this had considerable benefits for the wireless industry. He published his findings in 1905.

Enter the American Lee de Forest. De Forest was born in 1873 to an unusual family. His father was a stern, humorless man with a rather terrifying face. A Congregational minister, the father became president of Talladega College for Negroes east of Birmingham, Alabama, in 1881. Since the de Forests were white, their white neighbors considered them pariahs, so young Lee found himself alone quite often. He turned to science and invention. His prowess earned a scholarship to Yale, where he was known as the 'homeliest and nerviest student in school.' He earned a PhD in 1899, and formed his own wireless company.

De Forest was an odd man, fixated with finding his idealized 'Golden Girl.' Despite his unimposing physical features, he married four times (only three marriages were consummated – the first was a publicity stunt perpetrated by one of his business partners).[69*] He had the unfortunate

---

* His first wife, Lucile Sheardown, refused to consummate their marriage. During their honeymoon in Britain, he gave her a 'vigorous spanking,' and sent her home. He concluded that she was the mistress of a beer company executive and agreed to the marriage as cover. They quickly divorced.

habit of not giving credit to his creative predecessors, which got him into a world of legal trouble.

One of the problems intriguing him involved radio. Guglielmo Marconi invented wireless communications in 1895, but it was fit only for sending the dots and dashes of Morse code. De Forest wanted radio to send music and voice.

He read Fleming's paper and began his own experiments. Working with a small Fleming valve, he put a nickel plate beside the filament with a wire sticking through the top. He patented it as the 'audion,' a static valve for wireless, despite the fact it didn't do anything. On 25 November 1906, de Forest built a tube with three elements, a *triode*: a filament, a plate, and a nickel wire between them, placed as close to the filament as possible. At the suggestion of his assistant, John Grogan, he crumpled the wire to create a larger surface area, which he called a 'grid.' If the grid was positively charged, it attracted the electrons from the filament and funneled them down to the plate. This increased the number of electrons over what would have flowed without the grid.[69]

Moreover, the slightest variation in the positive charge to the grid was amplified in the flow of electrons to the plate. If de Forest slightly increased the positive charge a little, the flow of electrons increased far more. If he reduced the charge, the flow decreased considerably. The tube acted as an amplifier. De Forest patented the device in 1907 (still calling it an audion)[68] and a few years later used it to broadcast Enrico Caruso singing at the Metropolitan Opera. Unfortunately for de Forest, the public wasn't much interested in his audion just yet, and he sold the patent.

He eventually earned 300 patents in his lifetime, including one for the first sound movie in 1923, ahead of Edison. He also provided generations of work for attorneys, either suing or being sued for patent infringement (including a suit by Fleming over his second audion). He never made much money from his inventions, but his lawyers did.

If the public wasn't ready for radio, the telephone company certainly was ready for the audion. Engineers for AT&T bought the audion patent from de Forest, and vastly improved the efficiency of the tube by increasing the vacuum. The device enabled the Bell System, AT&T's telephone company, to amplify conversations over its wires. More important, the Bell System used millions of valves as switching devices in its telephone network. AT&T used them in radar during the war and later in microwave transmission. Valves became the backbone of the entire communications system of the US. It was a brittle, flimsy infra-structure, difficult and expensive to maintain and potentially unreliable.

De Forest valves – vacuum tubes – had several well-known weaknesses:

- The tubes were bulky. They had to enclose three or more elements (the filament, grid and plate) in a vacuum. Those elements had to be spaced apart so that the electrons flowed only when they were needed. The more complicated the tubes got, the larger they became and the larger became the device that used them.
- Tubes were expensive to build. Their many components had to fit together precisely. Mass producing them was never cheap, even with economy of scale.
- Tubes were fragile and failed regularly. You could not put them in an appliance or device that might get knocked around because the tubes would break or come loose. Even the tiniest vacuum leak ren-dered them useless, and they wore out regularly when the filament snapped.
- The tubes consumed a lot of energy. Ralph Bown of Bell Labs, lik-ened their use to 'sending a twelve-car freight train, locomotive and all, to carry a pound of butter.'[70]
- The tubes had to warm up. You would turn on a radio or a radio transmitter and wait several seconds, or minutes even, for the fila-ment to become warm enough to emit electrons.

Mervin Kelly was convinced as early as 1936 that vacuum tubes were a technological dead end. The telephone system stopped growing with the war (except for the military network), but with the war over and the economy likely to enjoy a post-war boom, the demands on the network in the US would be brutal. The technology wouldn't sustain those demands. Kelly believed that the future lay in solid state semiconduc-tors, which physicists were then just beginning to understand. He very plainly described his idea to Bill Shockley. Kelly probably foresaw a recti-fier made from the semiconductor copper oxide replacing mechanical switches in the system. It's not clear that he thought in terms of amplifi-ers as well.[20]

Shockley was of the new generation of physicists, grounded as he was in quantum physics, who understood what electrons did in solid con-ducting material. His understanding of solid state physics was one of the reasons Kelly hired him before the war, and why he turned again to Shockley when it ended.

Shockley thought he understood Kelly's goal. That was one reason he pursued his semiconductor work with Brattain before the war, and was what he hoped to investigate once the war was over. In January 1945, with Germany in ruins and Japan smoldering, Kelly asked the Pentagon if he could borrow Shockley back again, at least part-time.

$$\prec$$

Getting Shockley's attention wasn't easy. The government still had work for him, mostly using his analytical skills in assessing the dangers of the new world. Every week, he rode the trains between Washington, New York and Boston for the military.[71] Particularly worrisome to the Pentagon was the prospect of the Soviets getting the atomic bomb. The military asked Shockley and Jim Fisk to analyze how knowledge of the bomb might be discovered (without being let in on many secrets), and guess how long it would take the Soviets to duplicate the effort. The task involved several of Shockley's interests: the bomb, operations research and his growing fascination with the modes of creativity, particularly creativity and invention in groups.

Shockley was also asked by the state department to see if he could estimate how far behind the US the Soviets lagged in military technology.[89] He guessed at three years behind. Three years later, the Soviets exploded an atomic bomb of their own, helped, admittedly, by a productive spy at Los Alamos that surely didn't figure into Shockley's formula.[72]

Every month, Shockley hopped a train to sit on the five-man Joint Research and Development Board headed by Nobel Laureate I. I. Rabi. He got some lecturing assignments at Princeton, which enabled him to get his head back into physics. When he could, he popped into Bell Labs at Murray Hill for a refresher program Kelly created for him.

Bell raised his pay from $785 to $900 a month, enabling him and Jean to start thinking about buying a house.[92] Relations between Shockley and Jean were bad, however, exacerbated no doubt by his absences. In March 1946, they had a huge brawl over money, with Jean writing Shockley that she would no longer discuss budget matters with him without the presence of a disinterested party. She suggested he ask his bank for a budget counselor. The fight centered on his anger at the amount of money she was spending on clothing, especially her shoes and a coat, which she claimed had become shabby. She told him she'd buy whatever clothing she needed; she never told *him* what clothing to buy. She concluded: 'We cannot tolerate violence in this household. We are supposed to be adults.'[73]

It is not known what specific incident prompted that comment.

Instead of having it out with Shockley in person, Jean wrote him. Perhaps his punishing travel schedule made him a fast-moving target and that was the only way she could be sure to get her message to him quickly. More likely, she did not want to fight in front of Alison and Billy.

Shockley's temper was back. He rarely showed it at work; indeed his colleagues couldn't remember temper as a notable part of his personality. He apparently saved it for home. His violence to his children was not measured or rational; he acted in rage. He was more psychologically than physically abusive, but he could become suddenly infuriated and strike out, particularly at his son.

The boys took most of the burden of a failing marriage.

'[Abuse] was more frequently verbal,' Richard Shockley remembered. According to William: 'He hit me enough so I remember being beaten. It was never blows ringing down; it was always a few slaps on the back of my hand with a ruler.

'There was an issue of telling the truth. You always tell the truth. That was in the same category as don't run away when you're told to come. I was completely intimidated. You don't want ever not to tell the truth because you would be found out. One day I rode into the garage and broke the window on his MG with my bicycle. I went right in and I said, "Dad, sorry. I rode the bike into the garage and I went too close to your car." I figured that was the right thing to do rather than beating around the bush. We went out and took a look at it and then he slapped me. Who knows what would have happened if I had lied.'

The children looked increasingly toward Jean for protection but found none. Jean was 'overpowered' by Shockley, his oldest son remembered, and she seemed to her children, at least, to withdraw, to deny them support. 'She was sort of peace-oriented. He was kind of violence-oriented.' Jean, on the other hand, did not believe in confronting her husband in front of the children, apparently giving them the impression she was docile. She paid a heavy price for her restraint.

Shockley was increasingly bored with Jean. 'The reason was she was kind of mousy and boring,' a son said years later, 'and he was kind of volatile, traveled at 100 miles an hour. I knew that things weren't very good. I knew that because he'd blow up in the house a lot and he wouldn't come home a lot. He was away on business a bunch.'[74] Although, as one friend pointed out to Shockley, most people were 'mousy and boring.'

Nonetheless, on 6 September 1947 Jean had another son. They let little Billy pick the name – Richard Condit.

That year, they bought their first house, at 22 Academy Road, four blocks from the Madison train station. The house was (and is) a slightly eccentric beauty, three stories of vine-covered red brick (Shockley eventually pulled down the ivy) with a porch and a curved driveway. The house had a garage, a small back yard and an arbor. In the back was a screened porch to protect the family from the justly notorious New Jersey mosquitoes. Shockley used a second-floor bedroom as a study. There was a deck over the porch and two bathrooms. Above it all was an attic with another bathroom.

A particular joy for the children was the little private park. Academy Road, after a few hundred feet, became Academy Circle, a loop fronting half a dozen homes, including theirs. In the middle was a round grassy area spotted with ancient maples, a favorite place for the Shockley kids to play with their neighbors. Parents could watch them from the front windows.

Shockley's walk to the Madison train station took less than 15 minutes. After work, if there was light, he would lose himself in his garden.

Shockley had seen very little of his mother during the war, dropping in only when the Pentagon sent him to California. May couldn't travel east. The letters, telegrams and presents continued to flow between the coasts. The Shockley children thought of her as their rich grandmother back west. May, with little to do, complained to her son she was bored and depressed. Shockley suggested she get back to painting, and May, after a confidence crisis, began painting in earnest, mostly canvases with a strong Chinese motif, apparently inspired by her husband and the artifacts he collected, which still cluttered her Palo Alto apartment. She sold some of her paintings, though not enough to earn her living. Her investments were doing splendidly, however, and she was on her way to becoming modestly wealthy. William left her land, she had returned to the stock market after the crash, and she had rental properties.

Shockley, in his 36th year in 1946, had a splendid first house, a wife and three children, a steady income, a healthy solvent mother, and a rare portfolio of jobs, and he was at the top of his form intellectually. Within two years, he would be world famous.

Dark clouds gathered over the horizon.

—<

Mervin Kelly was the driving force. 'Nobody resisted Kelly,' Jim Fisk remembered.[11]

An intense, medium-sized man, a Missourian who majored in mining engineering as an undergraduate, Kelly was now executive vice president at Bell Telephone Laboratories. He had given up mining after a brutal summer in Utah and got a PhD in physics from the University of Chicago, working on Millikan's famed light experiment. Kelly was a firm believer in basic research and liked to run things his way. He was convinced that a semiconductor switch was necessary and inevitable, and that the best way to get one was to put together a multidisciplinary team of the brightest men, turn them loose on the problem and leave them alone. Kelly thought progress would come faster without pushing the research toward practical applications, an unusually enlightened philosophy. Shockley was just as goal-oriented. He wanted that switch and was content to put everything else aside. Shockley was a curious hybrid of the utilitarianism of Thomas Edison and the innocence of the pure scientist. He and Kelly were perfectly matched.

Shockley and Fisk knew exactly who would add theoretical strength to the team: a 38-year-old physicist and former petroleum engineer, John Bardeen.

Bell Labs at the end of the war was easily the largest and best industrial laboratory in the world, and was on its way to getting better. The labs already had won one Nobel Prize, Clint Davisson's. It employed 5,700 people, including 2,000 scientists and engineers. Money flooded in from AT&T and from AT&T's manufacturing arm, Western Electric division. Bell Labs did the research and Western Electric built the equipment based on that research. The new Murray Hill setup was among the earliest industrial research facilities to model itself on a college campus. Bell Labs could essentially buy any scientist it wanted. The labs felt a certain scruple about raiding other people's shops. That Bardeen was on leave to the Navy posed a problem for Kelly. Shockley and Fisk insisted.

John Bardeen was born in Madison, Wisconsin on 23 May 1908, one of five children in a middle-class, relatively functional family of considerable accomplishment. His father, Charles Russell Bardeen, was the first graduate of the Johns Hopkins University Medical School in Baltimore and one of the founders of the University of Wisconsin medical school in Madison. Bardeen's mother, Althea Harmer, was an expert on oriental art and practiced interior design.

Bardeen went to the University of Wisconsin, getting two degrees in electrical engineering, and in 1930 moved to Pittsburgh to work for the Gulf Research Laboratories studying the theoretical techniques for oil prospecting. Unhappy with his life as an engineer, he went back to school, to Princeton's illustrious physics department for the winter semester 1933. That was an unusual jump, and Bardeen must have shown something unusual in his tests and interviews. His mentor was Eugene Wigner and he was a student at the same time as Wigner's other protégé, Fred Seitz. (Robert Brattain, Walter's brother, and Conyers Herring were also students then.) He probably met Shockley and Fisk in their quick trips down from Cambridge during the MIT years.

Seitz and Wigner had done their classic work on the energy band structure of metallic sodium chloride, so Bardeen already had a strong background in crystals and surfaces. He impressed his classmates.

'He doled out his talents like precious nuggets in a seemingly parsimonious way, characteristic of his manner,' Fred Seitz wrote. 'He had many gifts, including a willingness to tackle complex problems with persistence and patience.' His only peculiarity was a passion for golf, unusual among scientists and academics, who generally consider it a grave waste of time and resources. He didn't play for the exercise or the fresh air; he was too competitive. He played golf to get the lowest score possible. Every hole was an opportunity for a hole in one.[75] He rarely spoke. Seitz was amazed when the two visited a fraternity house at the Carnegie Institute of Technology, where Bardeen boarded while working at Gulf Oil in Pittsburgh. The fraternity men were delighted to see him again and used his return as an excuse to throw a huge party. 'Without too much encouragement, he soon became the life and soul of the boisterous gathering,' Seitz remembered. Seitz also remembered going to bed long before Bardeen, who was still up partying.[75] Many men who knew him at Princeton also would have been astonished.

In 1935, Bardeen won a junior fellowship at Harvard in the Society of Fellows, an organization designed by the university as a community of scholars from different fields. Shockley and Fisk got to know him better there. Bardeen married in 1938 and moved to Minneapolis as assistant professor of physics, working under John Tate. He was particularly adept at merging quantum theory with solid state physics, publishing several notable papers. He also delved into superconductivity, a subject in which he excelled.

During the war, he went to work for Naval Ordnance in Washington on magnetic mines and torpedoes. He was invited to join the gathering at Los Alamos, but elected to stay with the Navy out of loyalty to his lab mates, Herring said. When the war ended, Bardeen was presumably going back to Minnesota until Bell Labs intervened.

Fisk told Kelly that he and Shockley wanted the 'strongest theoretical people we could find anywhere in the world,' and that meant Bardeen. 'Of course, Bill Shockley is one of the ablest people around anywhere and always had been,' Fisk said a few years later. 'But he recognized that we needed more and we decided jointly that there were probably only three people in the country that qualified here, and Johnny was probably the most penetrating of the lot.'[11] Shockley pushed the hardest.[76]

Kelly offered Bardeen a job and twice the money he was making at Minnesota. 'I could work on whatever theoretical problems I wanted to in connection with materials, so from that point of view it looked very good,' Bardeen said. Kelly also offered him all the money he needed, which made it look even better.[77]

By the beginning of the Second World War, physicists knew next to nothing about semiconductors. Semiconductors neither conduct electricity freely nor do they block it; they conduct grudgingly. Their conductivity varies with temperature and with purity. What scientists did know was thanks largely to the work of the British scientist Alan Wilson, who in 1931 put semiconductors (he may have been the first to use that word) in a quantum physics context. Some semiconductor devices had even been put to work, but only in limited ways.

The elements germanium and silicon are semiconductors. (So is carbon as a diamond, but no one thought of using that in any quantity.) Silicon is the second most abundant element in the Earth's crust (28%), after hydrogen. Think of sand, which is mostly silicon dioxide. It's also cheap. Germanium, a lustrous grey–white metal, was used in optical instruments. Both are found as crystals, although very rarely in a pure state. Germanium is much heavier than silicon, but works similarly; broadly speaking, anything said here about one applies to the other.

Germanium has four clouds of electrons around its nucleus; silicon has three. Both elements have four electrons swirling around in their outermost cloud. These are the electrons that matter most in semiconducting. Some of them have to move if the material is to conduct electricity at all. Metals conduct electricity well because they have lots of free electrons.

In a piece of silicon, each silicon atom shares one of its four outer electrons with a neighboring atom, so that four additional silicon atoms surround every silicon atom. This regular, repeating structure is called the crystal lattice. The framework is stable and sturdy and it requires large amounts of energy to dislodge any electrons. This stability can be disrupted with a trick now known as doping – adding a tiny bit of arsenic, for instance, to the pure silicon. Arsenic has five electrons in its outer ring.

Four of the arsenic electrons act just like the four silicon electrons in the lattice, but the fifth electron has no place to go. Shockley was fond of the analogy of a multi-level car park. If there are four parking spaces on the first floor and five cars enter, you have one car that needs to go someplace else. In a silicon crystal, those loose electrons can be aimed in one direction by applying an electric potential. Spare electrons (negatively charged) are repelled from the negative electrode and stream to the positive electrode. This flow of electrons, the current, passes through the crystal and the silicon becomes an efficient semiconductor. In the garage analogy, the car with no place to park will be aimed up the ramp to another level, and so is every other car that enters the filled-up ground floor. More accurately, if a new car enters, another parked car is shuttled upstairs so that the new car can replace it.

The situation can be reversed. What if we add a different element – boron – with three electrons on the outside orbit? Each silicon atom is now surrounded by seven electrons, four of its own and three of the neighboring boron. There remains a space for another electron, a hole, which is exactly what physicists call it.* Our analogous parking lot now has four empty spaces and three cars drive in. The three cars fill the first three parking spots leaving the fourth empty.

---

* Technically, it is an unoccupied energy state.

Adding electricity to a silicon–boron crystal doesn't have much effect on most of the atoms because electromagnetic ties hold most of the electrons strongly in place. The exception happens with any silicon atom abutting a boron atom. One of its electrons on the negative electrode side of the hole could get jostled into that hole when current is applied, leaving another empty space where it used to be. The new hole is one place nearer the negative electrode. Another electron slips into that space, leaving another behind and so forth. The holes 'move' closer to the negative side.

Arsenic-doped silicon and boron-doped silicon act alike, but physicists and engineers call them by different names. Crystals where the negatively charged electrons move toward the positive end (those with arsenic in this example) are known as negative or *n-type* semiconductors. Those with the holes moving toward the negative end (here doped with boron) are called positive or *p-type* semiconductors. In the n-type, the electrons are the 'majority' carrier and a simultaneous, lesser movement of the holes is the 'minority' carrier; the situation is reversed in p-type materials. The minority carrier was the key to what happened next.

Even before the war, semiconductors had limited use, the most famous being the 'cat's whiskers' in early crystal radios.

In the early 1930s, a lone, eccentric Polish-American inventor, Julius Lilienfeld, produced a crystal amplifier and applied for several patents. Unfortunately, the amplifiers didn't work very well (providing less than 10% amplification) and Lilienfeld lacked the resources to keep going. It's not even clear he made any that worked. His work was forgotten, but he got his revenge later.

Tube diodes couldn't store charges sufficiently; but a modern version of the cat's whisker, a small pellet of silicon and a tungsten wire, could.

Work on semiconductors went on during the war in several places, most notably at Bell Labs, the Radiation Laboratory at MIT and the University of Pennsylvania (Fred Seitz). But the most active center was at Purdue University in Indiana in the lab of a physicist with the strange and lovely name of Karl Lark-Horovitz.

Lark-Horovitz, an Austrian-born physical chemist, along with a half-dozen graduate students and researchers saved from the military, began work on germanium in 1942, with the theoretical assistance of Hans Bethe. Under contract with MIT, their goal was to produce effective devices for wartime radar. Lark-Horovitz reported in 1942 having created both p- and n-type semiconductors using a variety of doping materials, including boron, aluminum, gallium, indium, arsenic and bismuth. Because the products were so erratic, Lark-Horovitz's researchers had to greatly refine the process of producing their own crystals. His lab eventually produced large ingots of the materials, including a germanium ingot that could stand high voltage, and they carved detectors from the ingots.

'The irony of the whole thing is that two or three of us were occasionally around lunch asking ourselves, "Why can't we put a grid on this and make a triode of it to control the electrons?"' said one Purdue scientist, Randall Whaley. 'But in the press of getting degrees and putting detectors together for MIT, we didn't take the next step and try this.'[78]

No one knew more about the physics of germanium than Lark-Horovitz.

Research continued through the war. But in 1945, as it appeared obvious the war was ending, Lark-Horovitz and Purdue made a fateful decision. They decided that the lab would pull back from its emphasis on practical applications of germanium physics and concentrate on theory. Several of Lark-Horovitz's researchers continued looking at the possibilities, but the lab concentrated on basic rather than applied research. The decision made perfect sense: Purdue was a university, not an industrial lab, and had no customers for its products. It also had limited resources compared to Bell Labs or its cousins, Radio Corporation of America (RCA) or DuPont. With that decision, Lark-Horovitz probably walked away from immortality and a Nobel Prize, and Purdue from billions of dollars in patent royalties.[20]

On 17 April 1945, Bill Shockley again began puttering with a design for a field effect amplifier (putting a strong electrical field near a semiconductor to encourage the flow of electrons) and switch, a modification of his failed 1939 idea of using a p-n junction and silicon instead of copper oxide. 'It may be,' he wrote in his laboratory logbook, 'that the type of device considered here can be made of silicon with boron and phosphorous impurities.... There may obviously be very grave difficulties in applying these ideas in practical form but the nature of the process suggests very good possibilities in this direction.' Calculations indicated the device would work. Shockley had used what he came to call the 'try simplest cases' approach to devise the most pared-down design. It didn't work.

Finding out why Shockley's idea was wrong became Bell Labs' priority. By now, Shockley and a chemist were supervising a solid state physics group comprising 34 men. Shockley was directly in charge of the semiconductor research unit of the group. He had seven men, and he could – and did – draw from other members of the unit, or from other researchers at the labs when the need arose. The semiconductor group itself was interdisciplinary. Besides Pearson, Bardeen and Brattain, one of the cleverest and most useful was a physical chemist, Robert Gibney. Shockley and Morgan reported directly to Fisk, who was director of physical research. They had a half-million dollar-a-year budget, a huge amount of money in 1945.[79] Kelly remained at Bell Labs' New York office, determined to leave his Murray Hill team in peace, the perfect supervisor. The resources aimed at solving Shockley's mystery were impressive.

Several Bell Lab scientists, including Russell Ohl and Brattain, tried to produce Shockley's device. Everyone failed. Shockley even calculated how small the output of the device had to be for it to go undetected; the results didn't even reach that minuscule level. He estimated that he was getting 1500 times smaller effect than the calculations suggested. He had no idea why. The effort was a failure – it would become one of the most famous failures in the history of science.

Shockley made it a point of learning everything that had gone on during the war in semiconductors, which included a visit, with Morgan, to Lark-Horovitz at Purdue. Lark-Horovitz was out of town, but in the finest tradition of science, his graduate students opened his lab and showed them everything they had, including the new germanium crystal. MIT's RadLab also greeted the Bell researchers as old friends.[78] Kelly arranged for Ohl to demonstrate a radio he had built with crude solid state devices instead of tubes. The radio was highly unstable and no one, including Ohl, thought it would ever be useful. But there it was on a desk, blaring music without a tube in sight, a morale-boosting show-and-tell.

Bardeen joined the labs in October 1945, and Shockley was delighted to see him. He was assigned an office with Brattain and Pearson on the second floor at Murray Hill. Shockley's office was upstairs with the other supervisors.

One of the first things Shockley did was take his field effect design to Bardeen to see if the great theoretician could find anything wrong with it. Bardeen studied it for two weeks and decided that it looked fine to him. He later told Brattain he would have drawn a different conclusion from Shockley's theory, but the device still should have worked.[16] He was just as puzzled as Shockley and determined to find out what was happening.

—<

With the team in place and the work set out clearly, Shockley settled back into domesticity, and despite tensions, he and Jean resumed their social life.

Jean had no circle of her own friends. The woman closest to her was Cynthia Fisk, Jim's wife. For a while, Alison took piano lessons from Cynthia, whom she described 'one of the most cultured people I knew' despite being relatively uneducated. For a time, the Fisks lived a block from the Shockleys, who threw frequent dinner parties for them and Janet and Alan Holden (another researcher in the group) came. The Bardeens and the Brattains were often invited. Jean apparently was a very good hostess. 'Jean was a lovely, nice family woman,' Betty Sparks, Shockley's secretary, said. Usually, the talk was about their research, which must have driven the wives out of the room, but sometimes it was broader. Alison remembered a New Year's Eve party when the table discussion was solving the post-war world's problems. Everyone joined and she recalled it as a great evening.[8]

Visitors still were subject to Shockley's perverse sense of humor. He was famous for his collection of dirty limericks, and not all his jokes were kind or in good taste. John Pierce remembered one particularly unkind joke at the expense of Keren Brattain, a scientist in her own right who probably held her own in the heady scientific discussions.[12]

Alison, older than her brothers, seemed to have a different relationship with her parents – closer, somewhat more affectionate – but there was little touching, hugging or kissing.

At work, Shockley was a gentleman, his secretary Betty Sparks insisted, and for the first few post-war years, the team remained close-knit. They visited and chatted when they were not in the labs working. 'As a secretary,' she said, 'you know you'd get a feeling of dysfunction among the men who work there, but I never felt that.'[2]

Other Bell groups could borrow Shockley's men to help solve specific problems, marching into Bardeen's office for instance, and asking if he had the time to help them out. Shockley set the group's administrative informality and those asking for assistance didn't need Shockley's permission to borrow Bardeen; nor did Shockley object.[77]

Shockley, Morgan and Fisk held weekly seminars to stimulate cross-pollination between the disciplines and to make sure everyone knew what was happening in each other's labs. Having the physicists spend time at the seminars, at lunch, and at the bench, particularly with the chemists, proved useful, and the physicists soon grew especially appreciative of Gibney. 'We did more chemistry than quantum physics,' Bardeen recalled. 'We used essential ideas from quantum mechanics but not detailed theory.' Gibney was perfect as the ambassador from chemistry, and played an under-appreciated role in what followed.

At first, they had very little of significance to show for their work. The failure of Shockley's field effect ideas had everyone stumped.

John Bardeen had his epiphany on the afternoon of 19 March 1946.

After talking with Shockley the day before and that morning, he went to Brattain's blackboard and began scrawling diagrams. Brattain could only watch and nod. He told Bardeen he needed to think about it for a while, later called him back into the lab, this time going to the blackboard himself. Let me see if I follow you, he told Bardeen. 'Now Walter couldn't carry Bardeen's jock strap as far as being a mathematician,' his brother, Robert said, 'but he wasn't bad or stupid.' Bardeen agreed they understood each other exactly. Bardeen knew what was wrong with Shockley's field effect device.

Shockley had assumed that electrons drawn to the surface of the crystals were just as free as other electrons to move about. What Bardeen proposed was that they were not; they were trapped at the surface, blocking further current from ever getting into the heart of the crystal. The crystal was not responding to the current because the electrons had formed a barrier and few were getting past what was called the 'surface state.' Bardeen estimated that as many as a trillion electrons were getting trapped for each square centimeter of surface, and that only one extra electron per 100 to 1,000 surface atoms would be enough to block Shockley's effect.

That not only explained why Shockley's field effect didn't work, it explained a whole slew of confounding phenomenon the men had discovered.[23] Shockley must have been pleased and appalled: pleased because Bardeen had solved the problem, but appalled because not only had Shockley written on the surface state a year before, but he had published a paper on it while at MIT. The surface states in his paper then were largely mathematical and theoretical,[80] but still he must have wondered why he didn't think it would happen in the real world, in his lab. Within days Shockley and Pearson produced a field effect, but it was small and restricted to very low temperatures. The electrons were not as mobile as calculations predicted.[78]

To build a solid state amplifier, the researchers realized they were going to have to abandon Shockley's field effect and figure out a way to breach the surface barrier and deal with the plodding electrons.[22] They were back to theory.

Laboratory logbooks are a science historian's best tools. They rarely lie. Researchers put their most intimate scientific thoughts into them, and should they discover anything important they call in witnesses whose signature verifies that they bore witness to the matter. Every logbook at Bell Labs was numbered and the lab knew who got which book. The reasons were not just diligence or curiosity; the books were invaluable for patent attorneys. They traced inventions from their inception to their 'reduction to practice,' the point at which the lawyers could file patent applications. Patents, not science, were the *raison d'être* of the labs and the logbooks constituted the primary source of information and evidence that the work was done.

Shockley later used logbooks for another reason. Fascinated by the creative and innovative process, he went back to the logbooks of Brattain, Bardeen and himself to chart the number of entries and show how the number suddenly jumped in the fall of 1947, peaking at what he later called the 'Magic Month,' (actually, five weeks) leading to the invention of the point-contact transistor.

The logbooks of Bell Labs Case Number 38139 are a gold mine.

Brattain and Bardeen threw themselves into the task of building on Bardeen's insight. Curiously, Shockley did not, at least not full-time. He was busy with other things, perhaps demoralized by Bardeen's bolt of lightning. This strange detachment would have dramatic repercussions. Mostly, he left them alone. He was distracted by new challenges, particularly research into the flow of electrons through an alkali, a research problem he encountered on a trip to Europe with Bardeen that summer to see what European labs were doing. He also learned the joys of working at home. If the neighbor children in the park in front of the house did not make too much noise, he got more work done there than in the lab.

That meant he spent less time in Murray Hill, and less time in the laboratory. Most of the work went on without him.

The two men, plus Pearson and Gibney, were essentially left alone trying to breach the surface barrier at room temperature. Devices that only worked at the temperature of liquid nitrogen weren't going to be very useful. The researchers now worked in a state Shockley called 'Will to Think,' a phrase he borrowed from Fermi. 'A competent thinker will be reluctant to commit himself to the effort that tedious and precise thinking demands – he will lack the "Will to Think" – unless he has the conviction that something worthwhile will be done with the result of his efforts,' Shockley wrote.[23] Bardeen's breakthrough provided that conviction for the team, Shockley thought, although perhaps not for Shockley himself. That the puzzle was fascinating and the ground fresh and that the scientists were having genuine fun trying to solve it were less compelling motives, he felt.

By April 1947, Brattain could demonstrate the surface barrier by shining a light on n-type silicon and measuring a change in voltage at room temperature. The light produced holes that were drawn up to the negatively charged surface. In May, he went to work on the effects of doping on silicon. In August, Shockley proposed that the only way to break through the barrier was with very high-energy electrons. He had Brattain sign that statement in his logbook, self-defense for the patent lawyers.

The number of entries in the logbooks, which had slowly been building, erupted in late October and early November, particularly in Brattain's. Brattain and Bardeen were forming a symbiotic relationship, one creative organism with Bardeen's brains and Brattain's hands. Gibney provided chemical insight. Brattain did most of the documentation. In this period, Shockley occasionally dropped in to see what was happening and make suggestions. After talking to Shockley, Bardeen suggested they switch from silicon to germanium because they could get more effective use out of that element, particularly if they used Lark-Horovitz's Purdue ingots.

His researchers began a series of dazzling experiments in which the failures sometimes proved more valuable then the successes. It must have seemed to Brattain and Bardeen that they were being inexorably drawn someplace; advance led to advance, every mistake told them something useful. Shockley called it the 'Creative-Failure Methodology.'

The Magic Month began on Monday 17 November 1947. Bob Gibney suggested that Brattain apply voltage between the metal plate and the semiconductor while both were immersed in an electrolyte, a conducting fluid. Shockley called it a 'breakthrough observation.'* This generated a strong electric field perpendicular to the silicon surface. The electrolyte provided release from the surface state. To Bardeen, this suggested that an 'inversion layer' existed beneath the surface state, piling up electrons. Bardeen's idea was a 'basic new insight about the science of semiconductor surfaces,' Shockley admitted.

From submerging the crystal in an electrolyte, they finally reduced the electrolyte to a drop at the point where the contacts met the crystal.

The taste of adrenaline must have been almost palpable in the labs' air. Weekends disappeared. Days stretched to 12 and 14 hours. Within a week, Bardeen, Brattain and Gibney had two devices they could eventually patent. Neither was valuable in itself, but they provided foundation blocks of later patents, common practice in commercial labs. Three days later Brattain and Gibney had a third patentable device, a field effect amplifier using an electrolyte. On the 23rd, Bardeen produced still another field effect device that he credited to the 'Shockley effect.' (It wouldn't work then either, but it set the stage for later devices: a 'creative failure,' Shockley called it). Now they entered the phase Shockley called 'Respect For The Scientific Aspects Of Practical Problems.'

On 21 November they produced a field effect in the liquid electrolyte. 'It was just a few cycles a second, not worth a damn,' Brattain said.[103] In a seven-page logbook entry dated Saturday 22 November, Bardeen noted that the experiments 'show definitely that it is possible to introduce an electrode or grid to control the flow of current in a semiconductor.'

The next day, Sunday, Bardeen turned his attention to the possibility of an amplifier. On Monday, Brattain witnessed new 'disclosures' toward that end. Shockley came into the lab the day before Thanksgiving to witness the entry as well. They had an amplifier that boosted current about 10%; but that was not enough to be useful and the frequency was too low.

---

* An electrolyte is a conductor in which current is carried by ions (atoms that have either lost or gained an electron) rather than by free electrons.

On Thursday 4 December, Brattain performed three experiments, bolstering some earlier ideas. The following Tuesday they switched to n-type germanium. Accidentally washing away the electrolyte, they noted, revealed an anodized surface left behind that conducted better than the original electrolyte. It meant that the gold plate deposited on the surface was directly touching the germanium; not insulated from it by a germanium oxide layer, which is what they were trying to produce – another lucky accident.

On 15 December, Brattain replaced the electrolyte with evaporated gold in two points close together on the crystal.[80] Bardeen believed that a small electric charge would inject holes (the minority carrier) into the semiconductor surface, greatly increasing the capacity of the crystal to carry current.[20] One point served as the grid, the other as the metal plate. They got a slight voltage amplification, but no power amplification.

On the afternoon of 16 December 1947, Bardeen, inspired by Lee de Forest's triode audion, suggested that the two points be as close together as possible on the germanium crystal. Brattain devised a structure in which gold foil was spread over a plastic triangle, which touched a crystal of n-type germanium. Using a razor blade he cut a piece of gold foil into two sections 0.04 centimeters (0.16 inches) apart: one was the grid, the other the plate. The electrodes were attached to those points and held in place by the plastic triangle.

All this required the dexterity of a brain surgeon: The whole device was less than half an inch long.

The gold-tipped plastic arrowhead pointed down at the surface of the crystal. Three sets of wires connected to the contraption. One of them – a spring made from a paper clip to press the plastic down on the surface of the crystal – was bent like a coat hanger; another coiled wire, almost bulb shaped, plugged directly into the plastic; and the third, a thin wire connection, went into the crystal. It was a crude, almost preposterous, setup, far too ugly to inspire a revolution, and just larger than a shoelace tip.

When the germanium is simply sitting there, almost no current passes between the two pointed wires. If the third wire introduces a tiny current, the resistance between the two point contacts virtually disappears and a much larger current can flow between the wires.

The two men stood at the lab table and watched for the first time what would eventually be called the transistor effect. The power gain was 4.5; the voltage went up by a factor of four. 'Current flowing in the

Figure 13 The first point-contact transistor.

forward direction from one contact influenced the current flowing in the reverse direction in a neighboring contact in such a way as to produce voltage amplification,' Bardeen wrote laconically later.[78] They repeated the experiment and it worked every time. They could adjust the power and alter the gain.

Even then, Bardeen and Brattain knew what they had accomplished was important. That night, Brattain, unable to restrain himself, told his car pool buddies he had participated in the most important experiment of his life. (The next day he called them and swore them to secrecy.)

Bardeen and Brattain talked on the phone later that evening. They had no doubts they had succeeded in controlling the current in a semiconductor in a useful, practical way. Brattain knew what they needed to do immediately.

'I think we better call Shockley.'

CHAPTER 6
'There's enough glory in this for everybody'

Bardeen and Brattain had one immediate minor but annoying problem: they didn't know what to call the uncomely contraption on their bench. Brattain, of the two the least able to keep calm, couldn't help blab what he had done to his friends, despite the corporate secrecy that surrounded the work. John Pierce, his friend and colleague, was walking by Brattain's office soon after the invention, when the physicist invited him in, shutting the door behind them.

Brattain explained the unnamed device he and Bardeen had constructed. Pierce immediately guessed they had a replacement for vacuum tubes. 'I thought right there at the time, if not, within hours, I thought vacuum tubes had trans*conductance*, transistors would have trans*resistance*,' Pierce recalled. 'There were resistors and inductors and other solid states, capacitors and *tors* seemed to occur in all sorts of electronic devices. From transresistance I coined *transistors*.'[12]

'That's it!' said Brattain.[22]*

Transistor it was – a point-contact transistor to be specific, because the action end of the device was where the points of electrodes contacted the crystal. Now they had a name for it; they needed to make a practical tool from it.

Shockley was 'quite excited' by the telephone call, Brattain later remembered. Shockley didn't quite remember it that way.

'Frankly, Bardeen and Brattain's point-contact transistor provoked conflicting emotions in me. My elation with the group's success was

---

* Bell Labs, being a bureaucracy, formed a committee to find a name. They ended up with 'transistor.'

balanced by the frustration of not being one of the inventors,' he admitted later.[103] 'I experienced some frustration that my personal efforts, started more than eight years before, had not resulted in a significant inventive contribution of my own.'[23]

The telephone call changed his life, challenging his balance, his ego, and his loyalties. Fred Seitz, his oldest friend, believed his personality then began a transformation, a narrowing, an intensifying, an unbalancing.[81]

Certainly, he listened to the news with considerable pride. Bardeen and Brattain clearly had produced a breakthrough that could fulfill the dream of a solid state answer to the clumsy old tubes. The advance came from his people, working in his lab, on his watch. His career at the labs and his reputation there could only be greatly enhanced. AT&T operated one of the world's largest and most profitable monopolies, and the company had excellent reasons to reward people who contributed to protecting that monopoly and sustaining those profits. Shockley had every reason to think he would be well treated.

On the other hand, the point-contact transistor was the invention of Bardeen and Brattain, with considerable help from Pearson and Gibney. Shockley's role was more that of a guiding consultant than of an active participant. He detached himself from the day-to-day interaction and participation without thinking of the eventual price, leaving his men alone to do the work. Now who would get the credit? Whose name would be on the patents to come? Would his contributions – and they were hardly minor, in his mind or in truth – be recognized beyond the laboratory?

The role of the leader – Shockley – did not appear nearly as important as Shockley's theories of leadership and creativity demanded. Even Bell Labs management appeared none too happy with how the transistor came about. They worked very hard from the first to produce a mythology around the invention of the point-contact transistor that persists to this day – one of well-managed teamwork – that simply is not true. In fact, there was little evidence of closely directed teamwork in the invention.[79] Management's greatest contribution was to stay out of the way.

Shockley's contributions to the point-contact transistor were threefold. First, the work was based on many of his theories and completed by his people. Second, the breakthrough came from the failure of one of those theories. Third, he had the good sense to trust Bardeen and Brattain. By every standard, this constitutes a major contribution to the invention of the device. Apparently his involvement was too passive to provide Shockley with the credit he craved.

On balance, Bardeen and Brattain's success came as more of a blow than a cause for celebration.

At first, Shockley and his team resisted telling upper management officially of the invention, typical behavior in the labs, Brattain remembered. 'It was so damned important that they were scared if they told Kelly about it, it might be a flop. And they didn't want to advertise it until they were convinced themselves it wasn't a flop,' Brattain said.[28]

Excitement grew. Keeping the transistor a secret became impossible. Phones were ringing constantly and visitors suddenly found reasons for dropping in. Bardeen would go to the blackboard for anyone who asked, illustrating his explanations.[2] Everyone walking the halls knew something had happened and since everyone knew what Bardeen and Brattain were working on, they could guess.

After a few weeks and constant verification Shockley could wait no longer. He had to tell the top brass. His secretary, Betty Sparks, typed out a memo Shockley signed, 'Concerning the Report on Semi-Conductors,' inviting lab executives to a demonstration.

On 23 December, Harvey Fletcher, the director of physical research, and his boss, Ralph Bown, listened as Bardeen and Brattain displayed the transistor's ability to amplify speech through earphones 'in the tradition of Alexander Graham Bell's famous "Mr. Watson, come here, I want you!"' Shockley wrote. Amazingly, no one remembered or recorded what was said through the first transistor, but it worked.[23]

Bardeen's 23 December logbook entry was typically terse. 'Voltage amplification,' he wrote, 'was obtained with use of two gold electrodes on a specifically prepared Ge [germanium] surface. The Ge was high-back voltage n-type. The surface was anodized to produce p-type conductivity near the surface.' This entry was signed by $W=Shockley$ and other witnesses to certify what they had seen. Shockley had no entry for that day in his book.

History incorrectly records that event as the birth of the transistor age.

(In a great irony, had the original attempt by Brattain and Bardeen to prove Shockley's theory used silicon instead of germanium, the last experiment would have been unnecessary because the original experiment would have worked. The anodizing layer on germanium's surface complicated the physics. Silicon would not have produced such a layer. Had they used silicon, Shockley would have been right: his name would have been on the patent and the Nobel would have been unquestioned. They just didn't know that then.)

The lab closed for Christmas, and that night, at 3 a.m., the snow began. The storm didn't keep Shockley and some of his team from making their way to Murray Hill. The nearest train station, Summit, was not within walking distance of the labs, so presumably the men drove in, their car wheels wrapped in steel chains. Betty Sparks remembered that some even came on skis. How Shockley, Bardeen and Pearson made it in is not known. Brattain was not mentioned in the logbooks, and possibly could not get to the labs. While the snow reached for the second-floor windows, Pearson achieved a field effect at room temperature. Shockley witnessed it in Bardeen's logbook. Finally, when the storm got even more ominous, the men fled to their homes. The Blizzard of '47 ended 24 hours after it began with 25 inches of snow on the ground. Shockley decided to work at home the next day.

He never returned as part of his transistor research team.

Brattain, who often said the transistor group was the best research group in his experience, later sadly remembered Shockley's increased isolation as the moment of decline. 'The group began to break up because Shockley went off by himself and did his work,' he said.[28]

Underlying tensions contributed to the breakup. Bardeen was the informal leader of a group of theoretical physicists, including Herring and Phil Anderson (who would win the 1977 Nobel Prize for his work on computer memory). They worked in adjoining offices and took tea together regularly. In every physical science there is a gulf between theoreticians and those with a more sublunary view of research. Theoreticians tend to look down on their worldly brethren; applied scientists admire their pencil-sucking fellows, but believe science has virtue beyond knowledge for its own sake – that it should answer practical questions, solve practical problems, do practical things.

Shockley, more of the latter kind of scientist than the former, never completely fit in with Bell's theoretical group. Perhaps his experiences during the war, where hypotheses had life-and-death implications, fashioned his outlook. The theorists wanted to know how semiconductors work for the sake of knowing, sufficient motives for their 'will to think.' Shockley, hardly less talented as a theorist, was interested in the theory mostly so that it would lead to a solid state valve.

'They would be interested in the rigor of some particular paper,' Morgan Sparks remembered. 'Shockley was interested in the first few terms of something but what really was important to him was what it meant. He was perfectly capable of doing all the mathematics of this group, but he didn't do it. He'd just do the part of it he thought was important to the physics of the problem. There was always this division between this group. They recognized him for what he was, that he wasn't really one of them.'

Brattain had increasing difficulties tolerating Shockley, sometimes drawing his sharp cowboy language as a weapon from behind Shockley's back.[2] Bardeen had a temper and could hold a grudge. Shockley was obsessed by two ideas. One was his fear of being left out of credit.

No sooner had the Christmas Eve demonstration concluded than the Bell Labs patent machinery kicked into gear. Bown called Kelly, who called in the legal troops. AT&T kept two regiments of patent lawyers, one at headquarters in New York, the other in Murray Hill. They may have been the best in the world. Patents were how AT&T protected its monopoly. Every employee at the lab signed a statement as a condition of employment, assigning all patents to the corporation. They received $1 for signing, what lawyers call 'consideration,' to make the contract binding.[23] Lawyers began showing up to interview the participants to see who was responsible for the point-contact transistor and whose names should appear on the patents. Attorney Harry Hart interviewed Bardeen and Brattain separately about the other's contribution. Both men made it clear: the point-contact transistor was a joint effort by the two of them. Shockley knew that's what they would say.

Brattain recalled: 'He called both Bardeen and I in separately, shortly after the demonstration, and told us that sometimes the people who do the work don't get credit for it,' Brattain said. 'He thought then that he could write a patent, starting with the field effect, on the whole damn thing.' In other words, he could undermine their patent application. Bardeen and Brattain were stunned.

'I told him, "Oh hell, Shockley, there's enough glory in this for everybody,"' Brattain said.[28] Bardeen held his tongue.

Shockley's other, more productive, obsession had been percolating in his mind for several months, a different kind of transistor, one perhaps more useful and lucrative – and one that would reestablish his primacy. The point-contact transistor was impressive, sure, but he was unconvinced of the serviceability of Bardeen and Brattain's cobbled device, with wires sticking out and the whole thing held together by wax and a spring. Clearly, it would be difficult to manufacture and of only limited utility outside tightly controlled environments. Shockley had a better idea. His goal was to create a one-piece transistor, with all the physics packed into the crystal.

Shockley began the most fertile two years of his life, beginning with his own 'magic month.'[23] His 'Will to Think' now was fueled by fury.

The day after the storm, Shockley stayed home, drawing a brief outline of his plan for his logbook. As the logbooks themselves were never taken from the lab, he wrote on plain paper and later, using rubber cement, glued them into the lab books. On the 28th, he went into the office and had the logbook entry witnessed.

That afternoon, he took the Lackawanna Railroad into New York and then hopped on the New York Central's 20th Century Limited to Chicago for a series of scientific meetings.

After the first meeting, with a few days until the next, Shockley holed up alone in a room at the Bismarck Hotel, and on New Year's Eve began writing for his logbooks. It flowed and streamed and in two days he had written 19 pages on lined graph paper. 'With a suitable reduction in scale, it is possible to reproduce the conventional triode... structure using semiconductor in place of vacuum,' he wrote. This design was of little use – the manufacturing difficulties would be enormous and the device wouldn't amplify – but he had the basis for what followed. A second device written in the hotel room at the end of his stay was almost perfect, but he did not understand enough of the theory to know that.

The next day, 2 January, Shockley put his notes in an envelope and airmailed them back to Morgan, asking him to witness them and pass them on to Bardeen for his signature.

Shockley spent another day at the Bismarck filling in more sheets of paper with new ideas, mostly on the thickness of the semiconductors in his proposed device. He gave a talk at the University of Chicago, and then took a train to Cleveland to give another lecture at the Case Institute on nuclear diffusion, then on to Philadelphia for a visit with Seitz. He returned to New York on 9 January.

Meanwhile, back in Murray Hill, Bardeen continued working on ways to improve the germanium in the point-contact transistor. All research was halted by the New Year holiday and then by a vicious ice storm that forced the lab to close for another day. Morgan received Shockley's envelope on the 5th and signed it, and then passed it downstairs to Bardeen, who signed the entries on the 8th. The pages were then glued into Shockley's logbook.

—≺

Keeping all the devices secret at this stage was paramount. Many other labs, particularly at Purdue, continued pressing their research, and inevitably one would duplicate Bardeen's effort. There are times in science when a critical mass of knowledge leads to an inevitable explosion, and this was such a time. The danger became apparent during the American Physical Society meeting that Shockley attended in Chicago. Brattain, also in from New Jersey and recovering from a bout of the flu, saw Seymour Benzer and Ralph Bray, two of Lark-Horovitz's assistants, walking into a meeting room. Bray came over to chat. Purdue's researchers clearly understood the role of minority carriers in semiconductors, Brattain knew, which was the key to the solution of the puzzle. They just hadn't yet followed it to its logical end. At that time, Bardeen and Brattain were working on a paper for the premier physics journal *Physical Review Letters* on their discovery and feared someone would beat them to publication. That probably wouldn't affect the patents, but would be a serious loss of professional face. Bray wanted to talk business. Brattain, edgy, just listened.

'You know,' Bray finally said, 'I think if we could put down another point on the germanium surface and measured the potential around this point, that we might find out what was going on.'

Brattain gulped.

'Yes Bray,' he said, believing he had to respond with something intelligent. 'I think that would be probably a good experiment!' He walked away as quickly as politeness allowed. Bray had just described the experiment Bardeen and Brattain had performed a few weeks earlier – the one in which they invented the transistor.*

Working at a frenzied pace, Bardeen and Brattain got more than a 100-fold power gain on their high-voltage germanium on 16 January 1948. Three days later Brattain wrote in his logbook a 'disclosure' of using evaporated films of germanium as amplifiers.

---

* Bray and Lark-Horovitz apparently never did the experiment.

Shockley was racing on with his idea, stalking his house during the nights, pacing, tossing, turning. On 23 January, a Friday, still working mostly at home, Shockley discovered that the reason his earlier attempt at the new device had failed was that he too had ignored the role of minority carriers in the crystal, just as Bardeen and Brattain and Lark-Horovitz had in their earlier experiments. Indeed, Shockley decided that Bardeen's interpretation was not complete. Again using his 'try simplest cases' approach, he concluded that if holes were injected into the n-type material they would act as minority carriers, starting the cascade of electrons and holes. He suddenly understood how the device he imagined in the Chicago hotel room worked, and it was much simpler than what Bardeen and Brattain had come up with. He called the new device a 'High Power Large Area Semi-Conductor Valve,' which proved a serious misnomer: his device would excel at low power. He later named it the junction transistor. He still did not have a good handle on the theory.

'The device,' he wrote, 'employs at least three layers having different impurity contents.'

The result would be amplification.

Shockley said later that he had not been trying to invent an amplifier on the 23rd, but was trying to work out experiments for the point-contact transistor. The next morning he was up at 4:30 a.m. to write more information for possible patent applications. Still, he was so unimpressed that he waited until the following Tuesday to bring the notes in so they could be witnessed.[23] Only in the following days did he realize what he had.

In Shockley's conception, the works of a transistor were all subsumed into one crystal of germanium. During the refining process, the appropriate dopants would be added to the molecular structure of the semiconductor so that it had three layers. Contacts were attached to each layer. The two attached to the top and bottom layers, the n-layers, were called the emitter terminal and the collector terminal; the electrode attached to the middle p-layer was the base terminal. Batteries would be attached to the device to make the emitter negative relative to the base and the base negative relative to the collector. Current (the electrons) flowed easily from the emitter to the base, but flowed with difficulty from the base to the collector. Any change in voltage from the emitter to the base would produce an equal change in the current in their circuits. A change in the collector, however, would be greater going to the base – it would be amplified.

Figure 14 Official Bell Labs transistor publicity photo: Shockley (center) with John Bardeen (left) and Walter Brattain (right).

No one could actually produce such a device, however, because no one yet knew how to make germanium in that configuration, although several labs in the US (Purdue and Bell Labs especially) and in Europe (France particularly) were making long strides in producing crystals. Nonetheless, it soon dawned on Shockley that he had figured out the theory for a transistor based on the theory of p-n junctions, with both types of semiconductor up against each other, a device of far more use than the point-contact transistor.

Instead of running current between two relatively fragile contacts, as in the point-contact transistor, Shockley's idea was to have minute amounts of two dopants (amounts so low they would be hard to find chemically) in one piece of semiconductor, controlling the current flow between the junctions of the two. Since the junctions were much bigger than the contacts, Shockley's junction transistor could handle much greater currents.

Now, someone had to figure out how to build one.

Shockley, however, still had distractions.

The patent lawyers had four patents ready for the point-contact transistor. The first three were filed 26 February 1948. Patent one carried the name of Brattain and Gibney for their work using electrolytes. The second involved the inversion layer beneath the surface, and credited Bardeen. Gibney's name was on the third, also dealing with treating semiconductor surfaces with the electrolytes.[80] The fourth and key patent, the patent for the point-contact transistor itself, was being held up by the lawyers, in part because of the complexity of the problem, but in part because they needed to clear up any problems with credit.

They concluded that Shockley's name could not appear on any of the point-contact transistor patents. They could not name him without describing as his major contribution the theory behind field effect devices, and that ran right into poor Julius Lilienfeld, the lonely, mostly forgotten inventor who had taken out three patents on a field effect transistor in the late 1920s without actually producing the devices. Patent lawyers hate confusion (unless, of course, they need to deliberately produce it themselves) because it gives challengers ammunition. No one feared that Lilienfeld or his heirs could defeat the Bell patent; the lawyers just abhorred potential ambiguity.*

The labs even formed a committee to investigate Lilienfeld's patent, hoping he had either lied or erred, evidence for countering any future claims. The committee found that Lilienfeld did neither.** Shockley could not get patent credit for the point-contact transistor. He was furious. The fact that Bardeen was on the committee probably didn't help relations between the men.

Although there exist no records to prove it, sometime during this period, Shockley is believed to have threatened – perhaps indirectly – Bell Labs management with a challenge to the patents, claiming primacy on the underlying theory. He certainly made his displeasure clear to Kelly several times. Later events, especially management's subsequent attitude toward Shockley, strongly support that suggestion, and certainly it fit his mood.

---

* In later years, Lilienfeld's widow gave the American Physical Society money for an award for physics research, one of the biggest in the world. APS had to research quietly to find out who Julius Lilienfeld was. No one had heard of him.

** The Patent Office in fact first rejected two of the patents because of Lilienfeld's work. The crucial Bardeen–Brattain patent was not affected, nor was Shockley's later junction transistor patent.

Word came down from management that every publicity photograph of Bardeen and Brattain had to include Shockley. Shockley would also play a prominent role in the official announcement of the invention, planned for June 1948.[28]

As far as the world knew, the first transistor was the work of three men, John Bardeen and Walter Brattain and William Shockley, and that myth persists today, engraved in history by the Nobel committee and an oversimplifying media.

Bell Labs soon began a series of seminars to bring its scientists and engineers up to date on transistor development and to find ways of manufacturing and using the devices. At one, on 13 February, Bell physicist John Shrive reported a way of getting the emitter and collector points especially close together using a sliver of germanium and placing the points on opposite sides, which produced good transistor action. That was the tip-off to Shockley that the secret of the junction transistor lay in the minority carriers. He went to the blackboard and gave a blazing analysis of what Shive had proposed, making it clear to all in the room he had given the matter considerable thought himself, and had kept it secret. The display left everyone else, including Bardeen, speechless with admiration.[21,23,82]

Back in Indiana, Lark-Horovitz was still pouring considerable efforts into his semiconductor work. He wrote to Shockley asking for samples of Bell's n-type silicon. Bardeen and Brattain approved the shipment as a courtesy; Shockley objected. He did not want the Purdue team to muddy the patent waters. Lark-Horovitz, not getting the quick and courteous response he expected, wrote again. At the end of March 1948, Shockley told Lark-Horovitz he could not comply because it could involve patentable material, which must have sent a chill through the Purdue scientist.[83]

With Brattain busy trying to improve the point-contact transistor, Shockley turned to Morgan Sparks for help in trying to build a junction transistor. Sparks began filling the Brattain function – hands – while Shockley took over Bardeen's – brains. At first, Sparks got nowhere. One issue was trying to draw the crystal of germanium, all properly doped. Several scientists were working on that problem, and were finding the going tough, financially and technologically. Gordon Teal, whose research eventually led to the manufacturing breakthrough, kept putting in requests for funds to work on the problem and kept getting no response. Management apparently didn't think that pursuit worthwhile.

Another man, William Pfann, who later invented the method for producing the crystals economically, was actively dissuaded from his efforts
by management, including Shockley.[20] One scientist described handling
the dopants as 'like trying to separate a pinch of salt evenly distributed
through a trainload of sugar when you didn't even know the impurity
was salt.'[84]

Shockley had yet another problem. He prepared a paper for *Physical
Review Letters*; the editors said his explanation of the quantum theory
behind the p-n junction was insufficiently rigorous.[20] Shockley agreed,
and went back to the logbook. The extra work paid off eventually.

On 26 June 1948, Bell filed a patent application for Shockley's junction transistor, but only after Shockley apparently went shopping for his
own patent attorney at AT&T and found Rudolph Guenther.

The point-contact transistor solved many of the vacuum tube's problems, but not all. It used a very little electricity to do considerable work,
it did not require a warm-up period, and likely would be more reliable.
Unfortunately, it would be difficult and expensive to manufacture, and
while it could be made smaller than any vacuum tubes, there was a limit
to how much smaller. It was also fragile; slamming a laboratory door

Figure 15 The first junction transistor.

could make the point-contact transistor hiccup. It produced a lot of noise, some of which blocked its functioning.

If only Sparks could make one, Shockley's junction transistor would solve all the problems of vacuum tubes. It would use even less power, would have no warm-up period, and had the potential to be manufactured at microscopic size. Further, you could throw one against the wall and it would probably keep chirping. It would produce virtually no noise.

With all the pertaining patent applications filed, the labs felt ready to let the public know about the point-contact transistor. It scheduled a press conference for 30 June, two days after the junction transistor application, at its auditorium in West Street in Manhattan, convenient for the national press.

They had one worry. What if the military felt the transistor was of such great importance to national security that it needed to be classified? Since none of the work was done with government funds, the fear now seems baseless, but the Cold War was looming. The military was a huge customer of Bell Labs, and AT&T wanted nothing to upset that relationship; neither did it want the government interfering with the exploitation of the invention. Bell Labs sent Shockley, whose relationship with the military remained close, to brief the brass on the invention and derail any attempts to classify the transistor. Shockley found them fascinated but uninterested in interfering. He invited them to the press conference.

May flew in from San Francisco on the 29th to attend the Bell Labs party and to see her new grandson for the first time. She and Jean joined the wives of the other two men for a sumptuous dinner in the executive dinning room at AT&T.

On the morning of 30 June 1948, before the conference, executives from AT&T, Western Electric and the New York and New Jersey telephone companies enjoyed a private preview. Ralph Bown talked through a microphone handset to individual receivers in the audience, inserting and removing a transistor amplifier. He showed a waveform on an oscilloscope both with and without amplification, then did the same with a television set.[85]

Lark-Horovitz and several of his Purdue associates also received courtesy invitations to the public event. When he arrived, an agitated Seymour Benzer spotted Brattain taking his seat.

'What's all this about?' he asked. 'We've had some ideas about this.' He clearly was worried, and perhaps a little angry.

'Well, Benzer,' Brattain replied, trying hard to suppress a grin. 'I don't want to spoil the story. You listen and then you talk to me afterwards.'

Benzer paled as Bown demonstrated the transistor. 'No,' he admitted later. 'We had no idea of this.'[28] No idea, apparently, of what Bell Labs had accomplished, and no idea how close he and Lark-Horovitz had come.

Shy with the press and genuinely naïve enough not to worry about things such as credit, Bardeen and Brattain let Shockley answer most of the questions. He was much more at ease with public speaking and reporters. It furthered the public perception that the invention was a three-man achievement.

To the amazement of the labs people, the media were not much impressed by what they saw at the press conference. The newspapers blew the story. Even the *New York Times* downplayed the significance of the invention of the transistor, giving the story four-and-a-half inches in its radio column, just below the news that the soap opera 'The Better Half' had a new sponsor.[20]

The trade press too greeted the news with surprising unenthusiasm. Only *Engineering* and *Scientific American* had stories spelling out possible implications, both written by the same writer.

Confusion over credit soon seeped to the public domain. AT&T's press release, and most of the press credited Bardeen and Brattain working under Shockley, but not all. *Newsweek*, for example, named only Shockley, saying he 'came upon the principle while investigating the behavior of semiconductors.' The article, published three months after the press conference, didn't mention Bardeen and Brattain at all. Shockley, who had no interest in diminishing the importance of Bardeen's and Brattain's work – he only wanted himself included – circulated a note to the two men asking them what he ought to do about the *Newsweek* piece. 'Suit yourself,' Bardeen, wrote back. 'Walter and I are not worried about getting our share of the credit. If you do send it, we suggest you make it clear that Walter and I are from BTL [Bell Telephone Laboratories].'[87]

Shockley sent the letter and *Newsweek* published it.[88]

In public, at least, Bell Labs stuck to the company line – it was a three-man invention. But even within Bell Labs, confusion reigned, in part because of the insistence that Shockley get equal billing. No reporter asked why his name wasn't on the patent applications.

The version in the press release – that the three men deserve credit for the point-contact transistor – remains to this day the official version of the invention. And, every picture of the transistor's inventors, as ordered by management, shows three men.

Physical Review Letters editor John Tate accepted three papers from Shockley and broke all journal speed records getting them into print. The first two, both authored by Bardeen and Brattain, constitute the genesis of the electronic age. The third, by Shockley and Pearson, is largely forgotten. Gibney, by now working in New Mexico, apparently had become a non-person; his name is absent from the list of authors.*

While Shockley was working on the junction transistor, he was taking copious notes. Besides trying to satisfy the editors of *Physical Review Letters*, he decided to write a book. Bell Labs' publishing arm licensed the textbook to Van Nostrand in New York as part of the Bell Laboratories Series. Shockley had never written a book before and found it more work than he anticipated, because in part he had trouble 'freezing' the manuscript. Every time he had a better idea for improving his transistor, he ducked back into the labs to tinker and then back to the typewriter to revise the manuscript.[89] He wrote the first chapter last. Work on the book also was slowed by another business trip to Europe during the summer with Bardeen.

The Shockleys still socialized with the Brattains and Bardeens, although a certain coolness had set in. Now Morgan and Betty Sparks, Shockley's secretary, were added to the social group. Still working mostly at home, Shockley also kept track of Bardeen and Brattain by telephone.[89] Shockley put some of the mathematics of the transistor into a paper he wanted published before Purdue had a chance[90] and took the manuscript for it and the book with him on vacation with the Seitzs in Lake George.

The book, *Electrons and Holes in Semiconductors*, published in 1950, was Shockley's only book and became a classic of twentieth century science texts. In 551 pages, many of them crammed with formulae and graphs, mundanely bound inside a dull blue-gray and black dust cover,

---

* Many people believe, and not without reason, that Gibney was as deserving of the Nobel Prize as anyone, perhaps more so than Shockley. But the Nobels are limited to three people at one time.

Shockley captured everything then known about semiconductors, even pushing the knowledge about five years into the future with his detailed description of the junction transistor concept. Almost every electrical engineering class in the English-speaking world and most solid state physics classes used it either as a text or at least as a reference book. Bown supplied the preface. Shockley dedicated the first textbook of the electronic age to May.

All the while, Sparks worked at building a junction transistor. In Gordon Teal's crystal-making process, tiny pellets of dopants, about the size of BBs,* were added to melted germanium; then the mixture was drawn out as a thread of cooling crystal. The dopant atoms dispersed into the molecular structure of the germanium in tiny numbers, measured in parts per million atoms. Anyone analyzing the germanium might not even find the dopants, so little was needed. Sparks's pellets contained gallium (the p element) and antimony (n). The crystals were about a centimeter in diameter. The small size, the great virtue of what Shockley envisioned, wasn't optional. The minority carriers, in this case electrons, didn't live long enough to travel very far, so the smaller the crystal the better.

On 7 April 1949, Sparks and his assistant, Bob Mikulyak, watched a crystal slowly roll out of Teal's machine, turning as it came. When it cooled enough, they attached a direct current ammeter, touched slightly heated contacts to various parts of the little crystal and watched the meter record the differentials. The crystal had been transformed by the tiny pellets into a structure with p-n-p layers. The middle, n-layer, was extremely thin – eight to ten millionths of an inch (0.008–0.010 mils). They then went at the crystal with microblades to see how thin they could slice it and retain the junctions between the layers, etching and cleaning the faces as they went. They got a working crystal down well below 100 mils (a bit bigger than a grain of sand).

That night Sparks called Shockley to tell him that he and Mikulyak had built his junction transistor. Shockley admitted he had been wrong to discourage Teal.

Every transistor that powers the electronic age, the tens of millions now in our homes and offices, in our computers, watches, ovens, airplanes, CAT scan equipment, cars, fax machines, cameras, spaceships

---

* Tiny little pellets that are used in kids' target guns and air guns.

and yes, our telephones, is a descendant of *that* device. Shockley's feat – whatever the motivation – was his life's greatest accomplishment. It changed the world.

—<

The years 1947 through 1950, his most productive, would have been sufficient to make Bill Shockley one of the century's most important scientists. He could have retired right then and been world-famous, at least among physicists, and as his junction transistor worked its way into the lives of ordinary folk, his name would have spread. He never again reached that level of creativity or output.

Amazingly, he still had time for other things.

The Pentagon had work for him and kept a constant stream of classified and top secret papers flowing to his Bell Labs office. He brought in a safe just to handle the secret materials.[92]

Vannevar Bush and the military decided to gather 'the best thinkers' of the time to act as scientific consultants. He invited Shockley to join the group, which eventually became the National Science Foundation.[93] The council met several times and in 1947 agreed it was too small to handle the huge problems facing the country and recommended that 'an adequate, full-time section must be established within the framework of the Research & Development Board to reevaluate the general balance of the national military research and development program.' They would help decide which programs were worth pursuing and with what degree of urgency.[94]

Shockley's relationship with the Pentagon intensified, and eventually, in 1949, while he was on leave from Bell doing work with a large magnet at MIT, Shockley was offered a two-year contract as research director of the Weapons Systems Evaluation Group, replacing Phil Morse. He was very tempted. Partly, he was leading a tiring double life, working on the transistor and other things for Bell Labs as well as consulting for the Pentagon. The lab was increasing his administrative responsibilities, especially recruiting. When the offer came, he was commuting between Boston and Washington weekly. Shockley also felt he had a major stake in the success of Bush's group. It was partly his idea, derived from his research organization during the war. He was convinced that many of the techniques the group employed were relevant to civilian research. He was totally wedded to the idea that science should be practical to be worth doing. On the other hand, he didn't want to leave transistor development (at the time of the offer he had not yet built the junction transistor); it too had obvious defense consequences, which pleased his deep patriotism. In the end, to his wife's relief, he turned down the offer.

She told May she was afraid the tensions of the government job would give him an ulcer. She also wanted badly not to uproot the children, especially Billy, who was turning into a difficult child. Alison, now in high school, didn't want to move either, Jean wrote.[96] In the back of her mind she might have feared that someday she and the children would be alone.

What happened to Shockley next is what pop psychologists thirty years later would call a 'mid-life crisis.' Being Shockley, who knew no subtlety, it had a catastrophic influence on his life and on the world.

CHAPTER 7
'...To do my climbing by moonlight & unroped'

Shockley's transistor team began to break up the evening of 23 December 1947 when he went home in high dudgeon to work on his own project. No one lamented this breakup as keenly as Walter Brattain.

'I almost have a mystical feeling about the fact that the final discovery of the transistor, in a sense, waited for me,' he said years later. 'It was probably one of the greatest research teams ever pulled together on a problem.... I cannot overemphasize the rapport of this group. We would meet together to discuss things freely. I think many of us had ideas in these discussion groups, one person's remarks suggesting an idea to another. We went to the heart of many things during the existence of this group, and always when we got to the place where something needed to be done, experimental or theoretical, there was never any question as to who was the appropriate man in the group to do it.'[28]

Brattain pleaded with Shockley to keep the team together, despite the growing tensions. After talking to members of the group privately to probe grievances, Brattain wrote to Shockley to try (unsuccessfully) to patch things up. Shockley did not intend to rejoin his own team. He hadn't controlled it in 1947 and look what happened.

Shockley responded but only to himself. In the envelope containing Brattain's note, Shockley attached an undated, handwritten memo. 'I am overwhelmed by an irresistible temptation to do my climbing by moonlight & unroped. This is contrary to all my rock climbing teaching & does not mean poor training but only a strong-headedness.' He signed it $W = Shockley$, Madison, N.J.

Bardeen was unhappier even than Brattain. Shockley had cut him out of junction transistor research and was blocking his efforts in another field that had begun fascinating him: superconductivity, the effects on materials of extremely cold temperatures (approaching absolute zero, where all molecular motion reaches its lowest energy state). Further,

Bardeen did not like Bill Shockley. As early as 1949, Brattain warned Bown that Bardeen was thinking of leaving Bell Labs. Bown, apparently getting most of his information from Shockley, who was either oblivious to – or, more likely, insensitive to – the feelings of his underlings, did not take the threat seriously.

One Friday in 1950, Brattain and Bardeen walked into Fisk's office and demanded they no longer have to report to Shockley. The next Monday, they didn't. That did not solve their problem; Shockley still had considerable clout in the research done on their floor, even if he wasn't directly supervising the researchers. He would go from lab to lab, essentially butting into everyone's business. Bardeen seethed.

Six months later, in July 1951, Bardeen wrote a note to Kelly telling him he was leaving, unable to take it any longer and with a fine offer from the University of Illinois, made with the assistance of Fred Seitz, who had moved there. Bardeen charged that Shockley was interested in his working on projects 'only as he thought of problems of his own that he wanted investigated experimentally.... In short, he used the group largely to exploit his own ideas.' That put him in the position of competing with his own supervisor, 'an intolerable situation.' He told Kelly he was willing to talk about the matter some more[97] – a hint, perhaps, that he really didn't want to go.

Bardeen left for Illinois where he won a second Nobel Prize in 1972 for his superconductivity research, the work Shockley prevented him from completing at Bell Labs. He was the first person to win two Nobel Prizes in the same field, and only the second to win two, after Marie Curie, who won in physics and chemistry. His departure must have rattled Bell Labs management.

That left Brattain behind, still bristling. Deciding that Bell Labs 'didn't own my soul,' he went to Fisk and demanded a change in his status. What particular event or situation made him unhappy he never said, but he appeared to chafe at Shockley's undiminished power in the labs. Despite no longer technically reporting to Shockley, he still felt Shockley was interfering with his work. Fisk couldn't help him – either because he lacked the administrative power to do so or because he didn't want to take sides against his old friend. Fisk suggested Brattain call Kelly.

Brattain reminded Kelly that Shockley's junction transistor patent was vulnerable to attack, and if anyone challenged it, he and Bardeen would not help Bell Labs' position. A challenger could claim Shockley's patent was derivative from theirs. Clearly this event, three or four years after the transistor patents were filed, showed how troubled Bell Labs officials still were about patent challenges. Shockley was using the same threat against the Bardeen and Brattain point-contact patents to get his way. If other companies could use the Brattain and Bardeen patents to challenge Shockley's, the reverse could likely be true. All the patents were equally vulnerable. Kelly also knew neither of the two men would lie in a courtroom.

'This changed his whole attitude. And after that, my position in the laboratories was a little bit more satisfactory. I felt a little freer.'[28] Brattain was still not free of Shockley, however. Shockley still had considerable sway over what research was done and according to Brattain on at least one occasion was chased out of a lab by a supervisor who insisted that Shockley mind his own business.

Brattain left the labs in 1967, and ended his career teaching at his beloved Whitman College. Until then, the saving grace for him and those in the team unhappy with Shockley was that Shockley was busy elsewhere.

When the Korean War broke out in 1950, Bell Labs lent him back to the Pentagon. Shockley, assigned directly to the Frankfurt Arsenal in Philadelphia, centered his attention on developing a proximity fuse for the missiles that began appearing in the war.

He and Ed Bowles, his old buddy from the previous war, flew to Korea to evaluate operations.

Proximity fuses had been developed during the Second World War to use against enemy aircraft. They used a tiny radar transmitter to know when to blow up. Shockley thought the fuses would also work with artillery shells, set to explode overhead instead of on impact. The shells would blast deadly shrapnel over a much wider area. Even better, he suggested, they should make use of his junction transistor, which would be a lot better than relying on the miniature vacuum tubes in the shells used against Japanese suicide bombers in 1945. The military was excited by this prospect, and although those transistorized fuses never made it to Korea, the idea provided considerable encouragement to Sparks back in Murray Hill to improve the transistors.[2]

Shockley's most notable achievement in the Korean War was a flying loudspeaker.

Two truckloads of North Korean soldiers driving toward the Manchurian border in October 1950 were overflown by an unarmed C-47. As they reached for their rifles, a voice boomed from the plane demanding in Korean that they surrender, promising honorable treatment. The voice said fighter jets were in the area and if the Communist soldiers did not turn around, the planes would blow them to pieces. The trucks turned toward the United Nations lines, and other North Korean soldiers, as many as 300–500, came in from the hills, walking behind the trucks. Others removed the camouflage that had protected two other vehicles so they could be seen. The Communist soldiers all surrendered.

Could the loudspeaker be made large enough to cover 10 square miles, asked the Pentagon? Battlefields were complicated places and the broader the area covered, the more effective the system.[125]

Shockley went to Europe on Bell Labs business in the summer of 1950, writing May from the liner *DeGrosse* on the way to Plymouth that he was running laps around the deck before breakfast every day and had taken up fencing lessons on the ship.[126] He didn't tell anyone he and Brattain got into a shouting match on the boat, subject unknown.[98] From France, in August, he complained to his mother he had done very little climbing in the Alps because of the weather, and was embarrassed one day when a team of high school girls passed him on the way up to an 8,000 foot peak.[99]

His reputation outside the lab seemed unaffected by the problems within. In 1951, he and his colleagues received the John Scott Medal from the City of Philadelphia, one of the most prestigious if least-publicized science medals in America, and on 24 April 1951 he was elected to the National Academy of Sciences, the highest honor an American scientist can achieve – an honor that would later turn into sheer agony for both himself and the academy.[100] He also won the Air Force Association Citation of Honor, presented by General Jimmy Doolittle. He got the medal at a dinner in Los Angeles, bringing May down from Palo Alto.

—<

The point-contact transistor did not go into volume production until 1953; Bell Labs had that much difficulty trying to successfully manufacture the devices. Bell didn't announce Shockley's junction transistor until 1951. By then they fully understood the importance of Shockley's invention.

The same year, Bell Labs held a special symposium on the transistor at Murray Hill. The Pentagon asked the labs to invite 100 military contractors, and nearly 300 guests spent five days listening to lab scientists describe the devices. The proceedings were 'restricted' because of the potential military applications, and everyone present had to be a US citizen cleared by military intelligence and sworn to secrecy. Kelly told the group that point-contact transistors were now available in small quantities, and that the junction transistor would be out for experimental use the following year.[101]

The labs quickly published the proceedings of the symposium, also stamped restricted. Later, when the classification was lifted, it was published again and became legendary as 'Mother Bell's Cookbook,' the instruction manual for the transistor age.

By 1952, science writer Bob Cowan at the *Christian Science Monitor* could flatly predict the end of vacuum tube radios.[102] In 1953, *Fortune* published an article entitled 'The Year of the Transistor,' predicting that 'a pea-sized time bomb' was ready to replace the vacuum tubes. 'The new solid state devices will provide the reliability, compactness, and lower power consumption needed to lift information-handling and computing machines – the nub of the second industrial revolution now upon us to any imaginable degree of complexity,' *Fortune* wrote with surprising prescience.[103]

In fact, the general-purpose electronic computer had been invented just a year earlier than the transistor at the University of Pennsylvania by J. Presper Eckert and John Mauchly. By 1950, Eckert and Mauchly were already selling UNIVACS, electronic digital computers for businesses, behemoths crammed with tens of thousands of vacuum tubes. Every weakness in vacuum tube technology plagued computers with a vengeance. Eckert had designed ways to make the tubes more reliable (essentially by under-powering them) but his computers still took up entire rooms, used gigantic quantities of power, generated waves of heat, blew tubes regularly and were enormously difficult to build, essentially constructed by hand. To *Fortune*, the future lay clearly with a marriage between Eckert and Mauchly's machine and the little devices of Brattain, Bardeen and Shockley.

And so it eventually was. The marriage was the mid-20th century's cotton gin, the steam engine, fire, the wheel.

But by the time of the *Fortune* article, the devices had made little real impact. The several companies who manufactured point-contact transistors (Western Electric, Raytheon, General Electric, RCA) were producing only 50,000 a month, and Western Electric was sending most of its devices to the military. At the same time, companies – especially RCA – were producing 35 million vacuum tubes a month. Western Electric was constructing a plant in Laureldale, Pennsylvania that could turn out 1 million transistors a month. Meanwhile, they were hard to come by, being devilishly hard to make.

Raytheon was the biggest producer, with a tiny, unheard-of firm called Germanium Products of Jersey City, New Jersey, second. Twenty-five US firms and ten foreign nations took out licenses to manufacture transistors. At the end of 1952, Sonotone produced the first consumer product containing a transistor, a $229 hearing aid. A Germanium Products junction transistor replaced one of the three tubes in the device. A few days later Acousticon announced it had reduced the electronics to one transistor. No tubes.[103]

Bell Labs had a demonstration device, a two-stage amplifier, complete with resistors and condensers potted in a three-quarter inch cylinder of plastic. A faint sound fed into the device could be amplified to ear-splitting volume with a quiver of electricity. The engineers had packed another device the size of a book page that could do the work of 44 vacuum tubes.[70] The first transistor radio, the Regency TR1, was produced in 1954.

A grand future was coming.

Shockley, perhaps more than almost anyone else in the world, appreciated that future. He understood it much better than had Bardeen and Brattain. Even Bell Labs saw the transistor spinoffs as wizard aids to telephony. Shockley also began to get a glimmer of the financial possibilities. A person could make a great deal of money if he could control or at least direct the path the technology took. The word 'entrepreneurship' wasn't common in the early 1950s. Shockley was starting to understand it in a way few others did, and it began to prey on his mind.

He was also growing dissatisfied with his position at Bell Labs. Although management was bending backwards to keep him happy, they were not promoting him. The lab heads apparently knew his limitations as a manager, and other men, some hired after him, some his friends, were moving up to higher, better-paying positions. The company wanted him content and working, both for his unquestioned genius and, probably, because of the potential threat that an angry Shockley represented to their patents. So they increased his administrative duties, particularly with hiring new researchers.

His fascination with intelligence and the possibility of quantifying it and predicting creativity and performance grew. In his new role of chief recruiter, he could try out several ideas. One of Shockley's greatest talents was his ability to spot talent in others and make remarkable appointments, sometimes despite himself.

He also took over the task of determining merit raises at the labs. His files were filled with statistical analyses of pay data, drawing curves, sometimes department by department. In one case he thought the bell curve was skewed, so he cut the raises for three men.

He was even in charge of giving Brattain a raise to $1150 a month. The great Claude Shannon, creator of communications theory, went up to $1250 a month under Shockley's directive. Arthur Schawlow, aged 32, caught Shockley's eye and went from $710 to $775; Schawlow and his brother-in-law, Charles Townes, also at Bell Labs, would later win Nobel Prizes for their work with masers and lasers.

The reason Shockley was not getting promoted himself, unquestionably, was his limited people skills. He had angered too many colleagues. He had the reputation of being uncaring, insensitive and heartless, literally running a dozen men out of the labs when he decided they did not meet his exacting standards. Bardeen was not the only person to quit because of differences with Shockley, only the most illustrious.

It became increasing clear to Shockley his future at Bell Labs was limited. Fisk, who joined after he did, was already running the place and making more money than he was. Shockley thought that since he had actually invented something useful and Fisk, whom he liked and respected, was still only a bureaucrat, it should not be so. There was money and power elsewhere, he felt increasingly. That was only one of the many things on his mind. There was also home, Jean, and his marriage to consider.

Dick Shockley said his family just 'exploded.'

Shockley still spent considerable time on the road, between stints with the military and jaunts to Europe, the latter for months at a time, traveling by ocean liner (almost a week in either direction) and eventually by cloud-hugging prop-driven DC-4s. When he was home during the spring and summer, he spent hours in the garden, especially proud of the roses that bloom in blatant profusion in early June in New Jersey.

He remained attentive to his physical conditioning. He would come home from work at the labs, immediately head for his small gym in the basement and set up a chinning bar and jump rope while a phonograph blasted Offenbach's 'Gaiety Parisienne' endlessly. Jean got very tired of 'Gaiety Parisienne.'[98] He continued his rock climbing on weekends, and, when in Europe, mountain climbing. He often found beautiful young women to go with him. His favorite place was the Shawangunks in the Hudson River Valley, where he led the first assault on an intimidating cliff topped by a perilous overhang. After the climb, his team threw him a party at a nearby bar. The cliff still is known as 'Shockley's Ceiling.' He was part of an early generation of rock climbers still held in great esteem by those who scale cliffs for the hell of it.[104]

He was not a natural climber, said one of the beautiful young women, Marion Harvey, who ventured with Shockley in the Potomac Basin and West Virginia on Sundays when he was in Washington. 'Some people just flow,' she said, 'but Bill had to work at it.' While rock climbing now is highly complex, with a technology of its own, in the 1950s it was done with sneakers and ropes. No helmets or climbing boots with Vibram soles. Pitons were as complicated as the equipment got. Climbers would go one at a time, leader first. The others would watch from the base of the cliff. When the leader got to the top, he or she would belay the others in the group. There were falls, Harvey (now Marion Softky) said, but no one got hurt.[105]

Shockley grew more distant from his family, except possibly Alison.

Figure 16 Shockley and his sons, Dick and Bill (1959).

He considered himself an atheist and never went to church. Jean thought of herself as an Episcopalian but as she became unhappier, she became more religious and she briefly considered becoming a Roman Catholic. Alison was enrolled in the Sunday school of a Unitarian church because Jean liked the minister. Shockley often drove her and her friends to the school, regaling them with tales of his adventures, never talking about what they were doing at school. On the drive home he would ask no questions.

The family was characterized by its lack of physical affection. The few hugs and kisses there were were given with some embarrassment. May was like that and presumably also William Shockley senior, Alison said. So was her father, and so was her mother. It became a moat between the children and their parents. 'Neither of them were touchy,' Alison said. 'I can remember he liked to give gifts he thought were kind of surprises. He didn't like getting gifts because he didn't know how to accept them. Just to say "thank you," or "that was a nice thought" wasn't easy.'[98]

'They were not particularly good parents,' their daughter admits. Her brothers emphatically agree.

'I always thought of mother being much more passive. She did have a strong sense of principles and when she thought there was something that needed to be done that was right, she did it,' she said. 'I felt he was not able to give love and that had to do with his father and mother, maybe his genes. What comprises love, I don't know. He certainly did have a sense of responsibility. He was wonderful with me.' Her brothers did not have the same experiences.

One day Shockley was changing a light bulb in the garage and little Billy came to help. Shockley yelled at him, either because he was doing something dangerous or just getting in the way. The little boy ran back to the house crying. He got no sympathy from his mother and no hugs, and he remembered that incident decades later.

As Shockley's unhappiness grew, he spent more time with his daughter, sometimes traveling long distances to be home for her birthday, doing magic for her friends. Forty years later, her friends from high school could still remember his shows. 'He would take me around to friends houses and we would do card tricks.' He worked out coded arrangements with her so she could guess the cards, and she became a staple of his local magic act.

In 1952, Alison left home for Radcliffe. The following summer, Shockley took her with him to Europe. He had a physics conference at the University of Grenoble. She remembers the trip warmly: toward the end she was struck with an enormous crush on a student she met there. Her father arranged to leave the two of them together for a day.

The sons, born almost a decade after their sister, had it much harder.

'There was an incident where he found my football in the flower garden,' Bill recalled. 'I guess he liked the garden. Took it seriously. Apparently he had told me "don't play with the football in the yard because I don't want to go into the flower beds." He found it out there. He came to the kitchen and found me and confronted me with his normal way of asking questions, the answers to which were extraordinarily incriminating – that was one of the things that he did to take away a lot of my self-esteem – he could ask questions like: "Didn't I say don't do this? Did you then do it?" At that point, he hit me. The first shot I blocked, he swung an open-hand slap at my face and I blocked it. Then I realized, oh, I can't do this; I'm not supposed to do that. So, I stood there and let him hit me another time.

'I was so intimidated by him that the idea of running away was completely impossible – you didn't want to do that. You see other kids, whose dads would be chasing them down the street and it was incomprehensible to me, how could they do that? Aren't they just going to go to hell when their father finally catches up to them?'[74]

Jean always felt the intellectual inferior to her husband, as, by most standards, she was. So was most everyone else on the planet.

On 3 February 1953, Jean was diagnosed with uterine cancer.

A local doctor, C. H. Berry, found a growth the size of a cauliflower protruding through her cervix. A biopsy discovered an adenocarcinoma. Four days later, her uterus was dilated and scraped. A small tab of radium (60 milligrams) was inserted and she was radiated with 4,000 millicurie hours. She was sent home. On 1 April she underwent major surgery, the removal of both her ovaries.[106]

Shockley was fascinated with the situation. Here was a science he knew nothing about. As soon as he began delving he discovered that the medical community was not unanimous about the treatment, a division between doctors (mostly surgeons) who thought surgery and maybe a little radiation could cure the disease, and doctors (mostly radiologists) who thought massive doses of radiation after surgery were necessary. They disagreed on how much and of what kind. Here was an intellectual debate worthy of a great mind and Shockley immediately immersed himself in everything written on the subject, taking over Jean's treatment and undoubtedly driving her doctors mad. He decided Berry was not a significant player in the field and fired him.

He used every ounce of his estimable influence to track down the world's leading experts on uterine cancer and adenocarcinoma. He found them in Washington, New York, Boston and Los Angeles. His new membership in the National Academy of Sciences gave him access to them all. In some cases, he wrote letters describing what he understood to be her condition, and often sent along her medical records.

As far as he was concerned, the radiologists won the argument. On 11 June he determined that Jean would go to New York for deep radiation treatment.

Bill Shockley, the great theoretical physicist had, in a matter of months, turned himself into an expert on uterine cancer – at least so he thought. It may be safely assumed the doctors had seen nothing like him before. Since Jean survived this bout of the disease, his decisions are hard to question.

How much this fascination and activity were an intellectual challenge and how much it was deep concern for his wife can't be known; but likely it was more of the former than the latter. While she was recovering from her cancer he announced he was leaving.

Forty years later, the men who were his friends then were still appalled and told the story with suppressed anger. One version claims he marched into Jean's hospital room to announce the separation, but we'll never know. Clearly, he told her while she was still under treatment from her ghastly and frightening experience, when she was at her most vulnerable. He probably did not time it deliberately to inflict pain; more likely, he wasn't thinking about her at all.

Why he did it then is also a mystery. He was not under any external pressure to ask for the divorce at that moment. There was no other woman; indeed there is no evidence he had been unfaithful. Nonetheless, that's when he told her.

It was a few months later, in the autumn, that Shockley began thinking of another woman.

Her name was Jeanine Roger, possibly from Rheims, France. They met in September or October of 1953 in Paris. He was serious enough to pen bad poems in mediocre French, written and rewritten on hotel stationary until he was satisfied. He wrote about destiny, smiles in the moonlight and memories. She wrote that she hoped they would be friends – and could he send her some hosiery from New Jersey? She sent a picture from Nice, a lovely blonde woman with light eyes, perhaps thirty.[107]

With Roger, however, Shockley did an extraordinary thing – he lied. He told her his name was William Bradford. She called his hotel using that name, which meant he had to tell the switchboard operator that he might get a call in either name. In France, they would understand. She wrote at least one letter to 'William Bradford,' in Newfoundland, New Jersey, using a general post office address in a small town near Summit. It was one of the few times in his adult life that Bill Shockley lied.

He never saw her again.[*]

Jean, meanwhile unaware of all this, moved a cot with metal springs up to Alison's room, and told Alison she could feel every spring in her back, but the room was more comfortable.

During 1954, Shockley was frequently on leave from Bell Labs and from Caltech – where he was on a year-long sabbatical – to work at the Pentagon at the same time. He traveled between California and Washington often, visiting home occasionally. He and Jean told Alison they were to split, but they apparently didn't tell their sons what had happened – dad was simply on the road longer. But the boys suspected.

Bill remembered asking his mother if they were going to get a divorce. Even at his young age he realized the two never seemed to spend time together alone. When Shockley was there, they never stayed up talking after the boys went to sleep, something Billy had heard on the radio that married people do. No, Jean said, they weren't.

'I think she was doing what she thought she should do. I think where that was at odds with what was actually going on inside of her, she subdued reality. She was a *Reader's Digest* reader, *National Geographic*, all

---

[*]   The following January, he wrote to an organization in France, possibly private investigators, asking them to research the names of all the men killed mountain climbing in the Alps in 1953. He doesn't say why, and whether this has anything to do with Janine Roger is unknown.

seemed to paint a kind of picture where God is looking over everyone. The bad guys are obviously bad but most everyone is really a good person with wonderful qualities that can be called to the surface whenever necessary. My feeling is that that's kind of what she thought [life] was supposed to be and that's what she decided it was going to be.'

She could hardly have been surprised.

One night that spring, 1954, he was invited to dinner at the home of his climbing buddy in Bethesda, Maryland, Joan Ascher, a nurse, and her fiancée, Phil Cardon. Shockley was supposed to be there the next evening, but changed the date because he had to give a paper. His friends had another guest that night, Emily (known as Emmy) Lanning, also a nurse. 'She wasn't trying to fix me up or anything,' Lanning remembered later. 'It was an inconvenience because we liked to talk together and her fiancée then was there and having dinner with us... I said that was all right, we could do it again another time.'

Shockley walked in and handed his hostess a bottle of Jack Daniels and bowed. He was introduced to Lanning, and then generally ignored her.

Thirty-nine, pleasantly plain with short hair and thick round glasses, Emmy was not used to being ignored and shrugged to herself it was going to be a wasted evening. Shockley walked around the apartment looking at rock climbing pictures. They all had a drink and sat down to dinner. The meal was steak tartare, which Lanning couldn't eat, so she got hers broiled. While the men were telling jokes, some of them ribald, Ascher and Lanning talked nursing. After a time Lanning, who was not interested in their joking, went out onto the apartment deck by herself in the pleasant evening air. Ascher sat in an armchair in the living room; Cardon sat behind her strumming a guitar. The sliding window was open so that Lanning could hear Shockley read aloud the paper he was preparing.

Shockley had been studying creativity in laboratories and was about to read a paper on who was productive and who not. Emmy, who had studied statistics and operations research as part of her specialization in psychiatric nursing, knew a bit about this. During his explanation he said something she didn't agree with – she doesn't remember what – and spoke up.

'How do you know so much about this?' he asked, paying attention to her for the first time.

'Well, people are my business.' He was talking about people, she said.

He joined her on the davenport on the deck and finished reading his paper so she could be included. She gave him a brisk critique, which, she said, he seemed to appreciate.

'He was nice to talk to, bright, that was obvious, and I enjoyed the fact that he listened to what I had to say about his paper. Once he found out that I knew something about it that was of interest to him, he listened carefully to what I had to say, and he made changes in the paper because of me,' she remembered.[108]

At the end, she told him she'd like to hear him deliver the paper the next day. They began to chat about each other and were at it until 3 a.m. until, exhausted, they had to leave. She offered him a ride in her new Ford two-door.

'Are you married?' she asked him.

'Yes, but I'm not working at it,' he said. 'Why do you ask?'

'The men I meet are married and want to play around, or are mixed up, don't know what end is up, or they are not very bright.'

'I'm separated,' Shockley said.

She dropped him off at the University Club on 16th Street, a seven-storey brick building right next to the Soviet embassy, his Washington home. He asked for her phone number and address and said he'd like to stay in contact. She said later she didn't expect that would happen.

The next day, after delivering the paper (with Lanning present), he flew back to Caltech; Lanning went off to deliver a 'shit lecture,' a task that greatly amused Shockley when she explained it to him.

'I told him students were having trouble with a patient who was throwing feces at them and this was very horrible for them,' she said. '[The students] came from general hospitals where people use bedpans.'

Back at Caltech, Shockley accepted a consulting job at the Pentagon so he could get back to Washington in the fall.

It was not love at first sight for Emmy Lanning, she remembered.[108]

It wasn't for Shockley either.

Like many men reassessing everything in mid-life, Shockley went in search of a woman to act as a kind of transitional object to ease the way to the outside. There was a female doctor in Los Angeles, and he spent several evenings in Pasadena with a woman named Ruth, almost surely the Ruth Ross he grew up with at Stanford. Jean remained back in New Jersey and he spent virtually no time home.

No one knew how unhappy – disturbed even – Shockley was at this point. He began, uncharacteristically, a form of diary, small spiral notebooks in which he wrote down his feelings, often in pencil, usually from the back of the book forward. His thoughts were disconnected, stream-of-consciousness. Some were about business or about life's trivia; some are incomprehensible now. Others show a man who is terribly upset.

> ... day or so ago thought of cuddle tendency at age 16–17... A couple of days ago dream of strong man who fell from cliff on old rope and my guilt as senior climber... Idea of setting world on fire, father proud... Imp. or lack of appreciation by bosses means what?... MBS in Santa Clara defeats suicide... coldness BTL Directors... imaginary playmate... why ran and be caught by Wm.... Bids for attention... Solitary high school... Miss Richmond... Desire to cuddle... Night of calling father... The tent, the tickets, the girl, no show, no refund... Tie to tree & untie [Feeling of extreme tension and release] 15 Mar 1955.*

He saved these notebooks in his archives, this uncharacteristic introspection.

Whatever other distractions he had, Emmy Lanning was clearly on his mind. 'The flowers started right away, within a week or so, two dozen white carnations,' she said. 'He sent them from California. He wrote about T. S. Elliot; he could quote long passages. One note he wrote about time past, time present and time future.' He would also quote long passages from Walt Whitman. She wrote back, with some shyness, apologizing for her hesitancy. He reacted with a combination of atypical tenderness and typical, well, Shockley.

> I have just reread some twenty-two letters from you [he wrote]. I read them, of course, both with feeling and analytically. Just for the fun let's make a plot – quite likely, you may not like this, I am not

---

* The MBS is clearly May, Santa Clara is a community south of Palo Alto; but 'defeats suicide' is a mystery.

sure, but as you know I count and do other cold things, it's the way I am.

He then drew a graph on the letter, plotting the average mail he received every ten days, showing a rapid and statistically significant increase. He wrote that her style got freer as time went on and the 'handwriting somewhat more cursive. There are a number of touching thoughts,' he noticed. He found himself opening his soul to her in a way unmatched in any of his other relationships, warning her of his dark side, admitting things he would never admit to anyone else.

> I hope it will all add up to something worthwhile for you. This doesn't mean regrets or lack of anticipation, but just another warning that the deepest pessimism and general lack of admiration for the human race and myself are probably with me for keeps. I don't want you to get hurt by being too hopeful you can do something about it. Funny thing is this has not stopped me from doing good work in the last 16 years and probably won't now.
>
> ...Usually, I am geared into something that I really think is worth pushing. This has not been true the last four months and I have been rather in search of new directions.[109]

Neither of them knew the ground rules.

She ended one letter: 'Best wishes with no strings attached.'[111] The salutation on one of his letters read: 'Dear well-equipped female with brains.'

In none of the 'love letters,' except when he was quoting a poet, did the word 'love' appear. A note at the bottom of a letter, quoting a poet, read: 'We all wish to be loved alone,' he wrote. 'I try to carry on the psychiatric definition of love and to avoid the luxury of hopelessness and loneliness.'[110]

Her letters to him, however, were full of passion and longing. She made it quite clear she wanted him, and was in love with him, although none of her letters use that word either. She assured him he need not fear loss of control in their relationship as long as he continued to treat her well as he had. He kept up a constant barrage of flowers and she loved it.[112,113] He drove back to Washington in his Jaguar with Alison, who had been out in California visiting, to work for the Weapons Systems Evaluation Group

(WSEG). He set up an apartment on Massachusetts Avenue. He immediately called Emmy.

'We went to very nice restaurants, Chinese and all sorts. One at Normandy Farms in Maryland where we could sit outdoors. They had wonderful popovers. French, Chinese, I eat everything and so did he. He liked to eat out and eat in; he would try all sorts of food. We liked martinis; he drank wine. I like bourbon; he had a Manhattan when I had bourbon. At lunch, sometimes, we'd have a martini with a sandwich or something.... So we had a drink or two, wine with dinner, moderate drinkers, we could take it or leave it.'[108]

They talked very little about his work. He never mentioned the transistor, only his job at the Pentagon and the graduate students he had at Caltech. He told her he was looking for another job, and was getting several offers. He clearly admired her brightness and quickness and her ability to say what she meant.

Shockley found his soul mate.

CHAPTER 8
'Well-equipped female with brains'

Emmy Lanning came from Upstate New York, the town of Cazenovia near Syracuse, where her father ran an oil refinery. She had a younger sister, also a nurse.

Before starting nursing, Lanning graduated from Central City Business College in Syracuse and worked for a while in a local company. She enrolled at the St Lawrence School of Nursing near Lake Ontario, and then went to New York City's huge Bellevue hospital for general training. At Bellevue, she became absorbed by the psychiatric wards, a different kind of nursing than the physically oriented practice in the general wards. She went to Columbia University for some basic science courses and decided she was more interested in teaching nursing than in doing it. She finished her bachelor's degree at Catholic University in Washington, started on her master's, and began teaching psychiatric nursing at St Elizabeth's in Washington. Eventually, she got a job at Chestnut Lodge, a private mental hospital in Rockville, Maryland, which specialized in seriously schizophrenic patients.

She became interested in an analytical form of psychotherapy started by Harris Sullivan who was a student of the Harvard physicist and philosopher Percy Bridgman. Bridgman's contribution added an operations research slant to Sullivan's psychotherapy. 'Sullivan was very interested in looking into these interactions between patients. All of his work is based on the transactions that go on between people,' she said.

Lanning built the teaching program at Chestnut Lodge, beginning with six students drawn from one hospital and ending with 50 from five or six. The treatment, based on Sullivan's interpersonal methodology, had some successes. Her program became one of the best-known psychiatric nursing programs in the country, and she coauthored a textbook on psychiatric nursing, *The Nurse and the Mental Patient*. One of her students, down from Columbia, was Joan Ascher, who also wanted to teach.

Lanning's reputation was widespread, and in the fall of 1954, Mildred Newton, dean of nursing at Ohio State University, contacted her to see if she wanted to come to Columbus and start her own program. The offer was irresistible. Newton told her she could run the program any way she wanted. Lanning could not say no. She found an apartment in Columbus. That was the situation when, at Joan Ascher's dinner party, she met Bill Shockley. She said nothing to him about the move. They saw each other regularly, or at least when his travel allowed.

On one of their early dates, he asked her if she had ever heard of the transistor. She said, no, it was about physics and she really didn't want to hear about physics. 'I want to hear about you,' she said. 'I think he was taken aback.'

'We talked about him, and us, then he told me about his looking around for money and wanting to start his own laboratory and [start] recruiting people. I think that's another reason he took a special interest in my criticism of his paper, my knowledge about people.'

Finally, she had to tell him she was going to Ohio.

'I told him I was leaving the first of January, and he said, "really? You didn't discuss this with me. How long had I known?"

'I could have told him,' she said later, 'and I wanted to but didn't and this is why: "I could not place this kind of burden on you. This is my career and I love dating you but this is my career." I went on to explain what I wanted to do, told him the whole story, told him this job was the chance to do what I had wanted to do for so long. I have complete freedom in the program.'[108]

She left for Ohio the first of December. He remained in Washington.

Shockley began to sound out people who might know where he could either get a better job or, preferably, the capital he needed to start his own company. Kelly said he was sure Shockley wanted to make a million dollars, and offered to help him leave Bell Labs, possibly with hidden enthusiasm. His former commander in the Pentagon, Ed Bowles, acted as his agent. He talked to Vladimir Zworkin at RCA, just south in Princeton, and with the MIT administration. Yale offered him whatever he wanted.[110] So did Berkeley, but he was not interested. Clearly, whatever his reputation at Bell Labs, the rest of the physics community remained impressed. Nothing came of any of it.

Shockley also continued his role as Defense Department consultant. He was now in some demand as a speaker. In a speech to the Los Angeles Chamber of Commerce, he said that the days of the lone inventor were over and that teamwork now was required to accomplish anything, statements that were both wrong and disingenuous considering the history of the transistor. 'It takes many men in many fields of science, pooling their various talents, to funnel all the necessary research into the development of one new device.' He warned that the Soviet Union was far ahead in that kind of collaborative research. Sputnik was still two years away.[114]

Reporting on the transistor, he said that hearing aids were currently the largest single users of the devices, and that consumers would save $50 million because of them. Also, Pacific Bell had a $4.5 million switching machine using transistors that allowed operators to make long-distance calls directly without having to call another operator.

Shockley had begun feeling out prospective employers, apparently as a fallback position if he could not raise the capital for his own venture. His job survey was cut short by a spring trip to Europe on Bell Labs business. He kept up his correspondence with Lanning, telling her that Alison was going to get married to a man she met at Radcliffe, Gerry Ianelli. He approved.*

He was not as alone in Europe as he let on. He looked up the tall and beautiful climbing-buddy, Marion Harvey, who had been sent to Frankfurt by the government, essentially as a spy. She may have had a higher security clearance than Shockley. They spent the VE and Joan of Arc Day weekends together in Paris. She was flattered by his attention and it was clear to her that he was having a difficult personal life. He struggled to quantify the ineffable complexities of life, she remembered, intellectualizing feeling. He seemed more at ease when feelings were somehow made concrete, almost as if he were trying to find algorithms to describe his unhappiness.[105]

It was not an affair, Harvey said later. He told her about Lanning. Judging by her letters, which Shockley saved, they did not make love, although that was not, apparently, Shockley's choice.[115–118] When he returned to Washington, she and Shockley exchanged dozens of letters, at least one a week, for most of the summer. Then their correspondence

---

* Marry they did, and they remain so. They have no children.

ended. Although they would eventually live a mile apart in the last 20 years of his life, they never saw each other again.

—<

Shockley's crisis reached its apex early in the summer of 1955. In two months Shockley, aged 45, got a divorce, left his job, and decided to start a company of his own. He even sold his car, his beloved MG.

He wrote May, breaking the divorce news to her in mid-June. In early July, Jean drove to Maine to tell Billy. Shockley called his son and reported to May he did not 'seem too disturbed.'[119] May wrote back: 'Menopause for women and a similar change for men is a dangerous time. I hoped you both would be so occupied with your vital interests that you would weather the few years that come in the middle forties. That is all I have to say.' To Jean she offered her friendship and financial support.

Jean and May remained close correspondents for many years, chatting mostly by mail, sometimes by telephone, about home, the children, flowers, clothing. They visited each other several times and May began to treat her former daughter-in-law as one of her huge circle of friends. Her letters do not remain, but Jean's show the depth of their friendship. True to her word, May, who had learned to play the stock market like a pro, sent a constant stream of checks to Jean – unasked for – for at least ten years.

Like most divorcing parents, Shockley underestimated the impact on his younger children.

—<

The first person he told about his final decision to start a new company was Lanning. On 1 June he wrote her, explaining that he felt sure he would succeed because he was 'smarter, more energetic and [I] understand people better than most of these other folks.'[127] He wrote to May that 'lots of people [are] willing to back me up in a new venture to the tune of $500,000 plus in the next couple of years.'[128] Kelly arranged an introductory call to Lawrence Rockefeller, who could back a new startup company.[121] He and Rockefeller, however, never could strike a deal.

Shockley began traveling the country extensively, looking at what other companies in the nascent transistor industry were doing, companies such as Raytheon, Litton Industries, Texas Instruments, even Bulova Watch Co. He looked at whether any of them were making profits from transistors (they were not), and how they got their funding. He analyzed the cost of living in various places, including Southern California, the Washington area, Cambridge, and Michigan, and decided that costs did not make much difference. He read books on capitalism, and met with several industrialists such as William Hewlett of Hewlett-Packard, the biggest company on the San Francisco peninsula, and Edwin Land, developer of the Polaroid instant camera. Land expressed an interest in serving on Shockley's board.[129]

Sometime in the end of July, his diary contains a note to call the California industrialist Arnold Beckman. How he got Beckman's name is not known, but the two men met and agreed to terms in September: Beckman would fund a laboratory for Shockley as part of Beckman Industries. The announcement was made on 23 September 1955.[130] Shockley said he would put together 'the most creative team in the world for developing and producing transistors and other semiconductor devices.'

Shockley spread the word in the late spring and summer of that year he was looking for outstanding semiconductor scientists. Lanning said he did not recruit anyone from Bell Labs, feeling it would be unethical, but at least four lab men, including Morgan Sparks, came out to California to look around. Shockley was determined that he was going to get rich by leading the technology – direct the wave, not ride it. He scoured journals and the network of physicists he knew. On 10 October he jotted in his notebook some early work done on surface transistors by one Robert Noyce in Philadelphia, the first reference to the man who would be his most famous employee. Whatever else can be said about the enterprise, his ability to choose the best scientists in the field was unmatched.

Shockley's transformation was still incomplete.

He admired Emmy Lanning's quick intelligence and was beginning to trust her judgment too, especially about people. She became his chief interviewer. During recruiting, he would ask the interviewees to call Lanning in Ohio so she could talk to them. She asked general questions of the applicant ('Why do you want to work for Bill?'; 'What makes you think you'd like doing this?') and then she and Shockley talked over her impressions. He clearly respected her opinion, and he did not respect the opinions of many.

In October 1955 he came through Columbus on a recruiting trip. He had, she thought, been ready to talk about getting married before but she had deflected the conversation. Now he was divorced.* Shockley said he was ready.

'There are three issues as I see it,' Lanning remembers responding. 'I don't know if I can live with your brightness. You're too smart all the time. Second, no children. I'm forty but I never wanted children anyway. Three, if it doesn't work out, and I'm not sure it will, if either of us wants out, it would be just like that. No fighting, arguing or anything. Just call it quits.'

'I don't see a problem,' said Shockley.

Her being in Ohio was not deemed an issue. Her contract at Ohio State ran out in March.

'I love my job,' she said.

'I know. I don't care if you work or not.'

'Marriage is something I've never tried,' she said, 'and I don't know if I could make it work or not, but it's worth trying perhaps.'

'Let's get married Thanksgiving or Christmas. Do you want a church wedding?' asked Shockley.

Lanning said she had gone through a church wedding with her sister and wanted no part in one again, so they agreed on a civil ceremony. He asked if she wanted a ring and she said that would be nice.

Never in the conversation, marriage proposal and acceptance, was the word 'love' spoken, yet what resulted was one of the strongest of marriages, at least from Lanning's perspective, a true, great love story. 'He knew how to court. I liked it, and we had fun,' she said many years later.[108]

Her students in Columbus were intrigued. They insisted on meeting this most extraordinary man. On one of his trips, Shockley agreed to talk to a class. He told them about the transistor, giving his now-pat lay lecture. The students, however, were more interested in other things: why was he marrying Lanning?

Shockley joked several answers, but the students pressed him. He was not telling them what they wanted to hear. Finally, unable to escape, he said, 'Because she understands people better than anyone I know.'

Forty years later, she was still proud of that answer. Her students, who were probably more interested in answers that included 'because I love her,' probably were disappointed.[131]

---

* Jean had gone to Nevada, the easiest and quickest place to get a divorce then.

In Lanning, Shockley found a lover, his strongest, most loyal, most ferocious ally and the only true friend for the last third of his life.

For Lanning, the relationship was transforming. She totally submerged her persona in his. Ten years after his death, she could still weep for her 'sweetie,' still spoke of him much of the time in the present tense, still missed him profoundly.

A justice of the peace married them on 23 November 1955 in Columbus. The closest thing they had to a honeymoon was a quick trip to Chicago where Shockley had a meeting of the American Physical Society. Then he went back on the road and she returned to Columbus to break the news to her dean that she was leaving.

Shockley's children were surprised at their father's choice. 'I thought he would marry a glamorous woman and she's not,' said Alison. 'What he liked about her was her directness, her honesty, her intelligence. She could express herself, which neither dad nor mother could, in terms of feelings.... When dad introduced me to Emmy she came across to me very much a career woman and this is a woman who more than mother ever could, completely turned her whole life and being over to somebody else. I couldn't quite understand it.'

'I didn't like her too much when I knew her,' said Bill. 'I only really knew her during some summer vacations.... She just didn't really say anything to me; I wasn't impressed particularly favorably with her. I wasn't real turned off either. I like her a whole lot better now, the few meetings I've had with her after he died.'

Under no circumstances were his sons going to warm to her, no matter what she was like. She learned to be a neutral party to her husband's relationships with his children, which kept her out of the crossfire.

Their first year of marriage was an adventure.

Emmy flew out to California at Christmas time to be with Shockley, and arrived to the Pineapple Express, the chain of tropical storms – small monsoons, actually – that roar in from Hawaii and blast the California coast in winter. Shockley had train tickets to go to Palm Springs to meet Arnold Beckman, but the storms washed out the Southern Pacific tracks north of San Francisco, delaying the trip, so she and Shockley met May and her half-sister Lou Vee for dinner so everyone could be introduced. May and Lou Vee now were living together, in a state of passive aggression. They never really liked each other, but Lou Vee's husband had died and sharing expenses made sense. The two women spent most of the evening complaining about the weather. Coming from Upstate

Figure 17 Bill, Emmy and May at Rick's, 1969.

New York, Emmy was not impressed with the storm. A few days later she and Shockley made it to Palm Springs, one of only a few times she would meet Shockley's benefactor.

In March, 1956, she returned to Palo Alto to find a place for them to live. Shockley had taken a room with kitchenette in a motel on El Camino Real. He immediately became ill with a bacterial infection. His temperature soared to 104, deadly serious for an adult, and a doctor prescribed the right antibiotic in the wrong dose. A second doctor corrected the mistake just in time, and Shockley, now probably glad he married a nurse, was tended back to health in the motel room.

They sent Lou Vee, who had become a prosperous real estate agent in Palo Alto, looking for a house. Lou Vee sorted out the options from their requirements: Emmy had a lot of furniture back in Columbus and needed a house big enough to hold it all. Shockley wanted the house to be no more than five minutes from his lab. May recommended getting an apartment in Palo Alto to hold them until they found the house, but Emmy insisted she was ready for her own home as soon as possible. They paid three months rent on the now-unused apartment in Ohio rather than ship her things west and store them. Lou Vee found five possibilities within their price range that met the requirements. Emmy particularly liked one in Los Altos, just south of Palo Alto and west toward the mountains, and without telling Shockley which it was, insisted he visit all of Lou Vee's finds. He picked the same house she did, at 23466 Corte Via Road, where they would live for the next ten years.

The house needed very little work, Emmy said. It had two big bedrooms with a bath between, a pine-paneled den and an adjacent half-bath, a large living room with a fireplace, 'all very nice.' The spare bedroom would serve well when Dick or Bill came to visit.

Emmy looked for work and got some consulting jobs as far away as Sonoma, but it wasn't the same. She could not duplicate her job in California, even with a large veterans' hospital in Palo Alto. She learned to happily stay at home.[108]

In August, Shockley was appointed a lecturer in electrical engineering at Stanford, beginning a 25-year relationship.[122] His company was still forming, and in the autumn of that year he went to Europe to recruit, having found European scientists hesitant to take risks with a new company. It was not, he discovered, the European way.[132] He put ads in industry newspapers asking for engineers and metallurgists. He slowly began hiring.

Throughout the year, while still gathering his troops, he was full of research ideas. His diary bubbled with them, some he thought important enough to have witnessed.

This likely was the happiest year of his life – new opportunities, a new wife, a new home and his own company so he could be his own boss and do things without interference. The potential for wealth hovered over him. He knew that he could become fabulously wealthy. He was totally immune from the infirmity of self-doubt.

—<

Early in the morning of Thursday 1 November, a reporter from United Press called to tell him that he, Bardeen and Brattain had won the 1956 Nobel Prize for Physics.

He telephoned May after sunrise to tell her. 'Bill phoned me up. Nobel Prize,' she wrote in her diary. 'He thought it was a Halloween trick.'[133] In New Jersey, Jean found out when a reporter, thinking Shockley still lived there, called her to get a comment from him. She phoned Alison in New York. 'I cried. I was delighted,' Alison said.[8] Her brothers remember being less ecstatic.

Shockley spent most of the morning fielding telephone calls from reporters and friends. When things calmed down the next day, he gathered his forty or so employees into the restaurant at Ricky's, Palo Alto's landmark motel, for a celebratory lunch with champagne. He held court at the end of a long table wearing a huge grin. Even he was demonstrative, if somewhat stunned by what had happened. That night, he, Emmy and May went to Mings, then the best Chinese restaurant in the area, for dinner. His fortune cookie read: 'For better luck you have to wait till winter.' So much for fortune cookies.

That the transistor inventors were serious candidates was common knowledge for years. They knew they were finalists in 1954. The only question among Nobel watchers was whether Shockley would be included. Despite the secrecy imposed by the Nobel committee, rumors often fly from Stockholm before a prize, terrifying those named in the rumors, paralyzing them with anxiety. Sometimes the rumors proved wrong, creating incidents of writhing. Reporters once staked out the home of a 'sure bet,' only to slink away in the night when the announcement came that he had not won. The scientist, believing the rumors himself, was crushed and humiliated. He did win a few years later, but the people who called him first to tell him had a difficult time being believed.

Brattain called Bardeen to congratulate him and sent a telegram to Shockley.

Meanwhile, telegrams and telephone calls by the hundreds inundated the three. Bardeen at first professed a reluctance to go to Sweden, a trip that would intrude on his privacy and peace. He soon relented, and told Brattain he would be happy if the two traveled together. Work was out of the question; the tumult was so bad that Brattain found himself feeling lonely and ignored in those few moments when nothing was happening.[123]

No award in the world carries as much cachet. The story and picture of the winners appear on the front page of every newspaper. The words 'Nobel Prize winner' or 'Laureate' are placed within a word or two of winners' names as first reference in print forever after. Those names on petitions or grant proposals are sought endlessly, their opinions reported with authority. Everything they say publicly for the rest of their lives carries weight far beyond the convictions of mere mortals.

Shockley was incapable of embracing the honor unquestioningly. He was growing paranoid.

The Swedish Royal Academy of Sciences, which runs the awards, gets nominations from previous winners and from prominent scientists in each field. Brattain knew of one man who nominated just him and Bardeen, and another who put in all three names. The latter, apparently no friend of Shockley's, told Brattain he did not think any of them would get the prize unless Shockley's name was included. Davisson, a Nobelist at Bell Labs, also nominated all three.[28] Shockley must have had contacts at least as good as Brattain's and heard similar rumblings. The prizes are not above politics and Bell Labs and the American scientific establishment – which then embraced Shockley – thought the transistor deserved a Nobel and most probably concluded that Shockley had to be honored as well, but that was not unanimous. Pressure could be exerted.

Shockley heard rumors that his name had been opposed and made inquiries in Stockholm to verify the talk and perhaps see who had opposed him. The Academy wrote back that it was none of his business; all proceedings were secret and no one would divulge what happened. He was advised to enjoy the honor.

Alfred Nobel's will is quite general about who should get his award, leaving considerable flexibility to the various honoring institutions.

> ...the capital, invested in safe securities by my executors, shall constitute a fund, the interest on which shall be annually distributed in the form of prizes to those who, during the preceding year, shall have conferred the greatest benefit on mankind... one part to the person who shall have made the most important discovery or invention in the field of physics...

Most Nobels in physics go for theoretical work, the kind of blue-sky *Erleuchtung* work that makes a leap in knowledge likely to produce later practical changes in the human condition. Einstein's award for the photoelectric effect is a fine example: It was theoretical, but we wouldn't have television, navigation satellites or radar without it, to name just three applications. The transistor award also had both a theoretical breakthrough – Bardeen's surface state understanding and Brattain's point-contact transistor that demonstrated it – and an almost instant practical employment, Shockley's junction transistor.

Moreover, the point-contact transistor was invented by a team working under Shockley, basing their experiments on his understanding (partially erroneous) of how semiconductors worked, at a time when he knew more about semiconductors than anyone else alive, except possibly Lark-Horovitz. Bardeen and Brattain were spurred to invent the point-contact transistor by Shockley. Had Shockley been excluded, some could have argued that no great injustice was done, but they would mostly be people who did not like him. Giving Shockley a share in the Nobel Prize was entirely defensible; his claim was quite genuine.

How much his feeling of persecution spoiled the event for him is unclear; he left no record of his apprehensions other than the inquiries to Sweden. He was probably too busy to spend much time at it, and eventually the imperatives of Stockholm took control of his life.

The following weeks were full of arrangements, interviews, and shopping sprees. Shockley told Emmy to spend whatever she needed. She went up to the elegant City of Paris apparel store in San Francisco and bought a dress of rose-colored French silk with an Empire-style waist and short sleeves.

The three men agreed to meet in New York with their families and friends from the lab for a celebration dinner before setting off for Sweden. Invitations for their presence, social and business, poured in from all over Europe. May, who by this time billed herself as the 'grandmother of the transistor,' announced she was going too. She did not intend to miss her boy's greatest moment. Shockley and Emmy talked it over and concluded they ought to take her.

The dinner in New York was a glorious success. Bardeen's son came down from Harvard, and men like Pearson came in from the labs. Several grad students and former assistants of the men also came with their wives. The men received a loud standing ovation when they entered the room. Brattain admitted he was on the verge of tears much of the evening. 'An honest man cannot but admit that the acclaim of his closest associates is sweet music indeed, a very high spot, if not the highest of all.'[123] For one evening, the animosities were eased by a warm balm of satisfaction, friendship and applause.

The Bardeens, John and Jane, and the Brattains, Keren, Walter and their son Bill, left the next day on the 3:30 p.m. SAS DC-7 for Copenhagen, a flight well-lubricated with vintage champagne; the three Shockleys were to follow two days later, flying directly to Stockholm.

At midnight, New York time, the Bardeens and the Brattains arrived in Copenhagen.*

---

* Brattain left a marvelous 62-page diary of their adventures at the Nobel ceremony, which is relied on here.

It was a dull, rainy day, so they decided on sleep for a few hours at the hotel, then got up, made a few telephone calls, and went to the Seven Little Houses restaurant for dinner so that Bardeen could explore the famous wine cellar. Then back to bed – but they were awake after a few hours, enervated by the time difference.

The next day Niels Bohr invited them to a seminar at his institute. The following morning, they took a private tour of the Rosenborg castle, home of the Danish royal family. They found that being Nobelists-to-be had extraordinary influence wherever they went, even in Denmark. Museum closed? No problem. Officials simply opened one at the convenience of the wives.

On 5 December they left by plane for Göteborg, Sweden, an hour away, to be entertained by a former Princeton graduate student Bardeen knew. He took them on a tour of his university, led them to a rehearsal of Prokofiev by the Göteborg Symphony Orchestra, and fed them 'the works, with the starting toasts and responses. Some hors d'oeuvres on one plate, then a fresh plate and more. If you don't finish your aquavit and beer in time, you lose it. One learns by experience,' Brattain wrote. Then, after-dinner drinks. Brattain thought perhaps the sea had been emptied for their meals.

The next day: Stockholm by train. Brattain, noting the latitude, remarked that it got dark right after lunch. They arrived in Stockholm at 6:20 p.m. and found the official greeting party waiting on the platform, including several scientists, a man from the US Embassy and an official from the Nobel Committee. The platform was also crowded with photographers, and flashbulbs sparkled in the otherwise gloomy station as they got off. Cars came. There was a bus for the luggage. The parties were convoyed to the Grand Hotel. They checked in and immediately went to a press conference.

The Swedish government had selected a 'gofer' for each family – Brattain called them 'attendants.' A Mr Kjellberg took care of the Bardeens, while Gunnar Lorentzon was assigned to the Brattains. Kjellberg took them off to dinner after the news conference.

Brattain and Bardeen had adjoining rooms, 415 and 416, facing the inlet of the sea, across from the royal palace. Little steamers tied up at the wharf below, were loaded up every morning and left for ports unknown. Brattain suppressed an occasional desire to get on one and flee Stockholm.

The next day was filled with shopping and a reception at the Swedish telephone company office. Sometime about then Bardeen and Brattain might have wondered where the Shockleys were.

Back in New York, Shockley, Emmy and May arrived at the airport to find that their SAS flight to Stockholm had been delayed by mechanical difficulties. Anxious to get going, Shockley proceeded to outsmart himself. He booked his party on an Air France Constellation to Paris. Unfortunately, Paris was fogbound, so they landed instead at Bordeaux. They took the night train to Paris, spent the day there, and flew to Stockholm. Instead of being five hours late, they were twenty-four, arriving exhausted shortly after the telephone company reception. They too were housed at the Grand.

Shockley's arrival did not change the informal social structure that had evolved. The Bardeens and Brattains stuck together, spending time with the Shockleys only when necessary. A pleasant dinner together in New York was one thing; days together, apparently, were more than the men could tolerate.

The ceremonies that year were being held under a somber cloud. The Soviets had invaded Hungary a month earlier and were occupying Budapest, suppressing a popular uprising against the Communist government. The world still was reeling from the Suez War, which had ended two months earlier. The literature prizewinner, Juan Ramón Jiménez, the modernist Spanish poet, stayed home, mourning the death of his wife.[*]

Shockley and his colleagues met the other winners at the reception: Dickinson Richards of the US, André Cournand, a French-born American, and Werner Forssman of Germany who shared the award for medicine; and Nikolai Nikolayevich Semenov of the USSR and Sir Cyril Hinshelwood of the UK, the chemistry winners. Semenov and his wife were accompanied everywhere, even to the opera, by three men from the Soviet embassy – two always dressed in brown, one in black – obviously thugs to keep them from fleeing. The usually impassive Swedes, upset about the Soviets' action in central Europe, did their best to harass the guards. They had already chased the Soviet ambassador to Sweden home, although he returned briefly for the Nobel ceremonies and would

---

[*] An extraordinarily violent year, there was no Peace Prize awarded. The prize for economics was not established until 1969.

flee again when they were over. They tried to exclude the three men from as many official receptions as they could and make their assignment as unpleasant as possible. Since the Semenovs had two sons back in Russia, the bodyguards were probably superfluous.

Monday 10 December was the big day. Shockley and the other winners assembled for a rehearsal in an anteroom and then were escorted to the auditorium down the aisles, as they would be in the real ceremony. A Nobel official described the schedule to them. The press was setting up its positions and watched while the men walked through the presentation ceremony, with the Nobel secretary standing in for the king.

At 3:30, dressed in formal attire, the three men left for the Stockholms Konserthus, each with his escort. They waited while the audience was seated and then the royal family arrived to a fanfare of trumpets.

The new Laureates were then escorted slowly in a double line into the auditorium to more fanfares, each line led by a student wearing a blue and yellow sash. They assembled in an inverted 'V' on the stage, the Laureates on the right, the official sponsors, and their escorts, on their left. As each reached his seat, he bowed to the royal family sitting in a row of chairs in front of the stage on the right. Behind them sat previous Laureates and members of the various academies and institutes that gave the prize. A symphony orchestra perched in a balcony to the rear of the stage. Under the silver pipes of the organ, the yellow and blue seal of Sweden hung directly above a bust of Alfred Nobel, the man who invented dynamite and created the prize to atone. Yellow chrysanthemums covered the front of the stage.

Emmy and May sat in the front row. Student ushers stood at every doorway. Around and above them tier after tier filled with Sweden's elite and guests, reached to the crystal-glittered ceiling, everyone but the children in the most formal of dress, the men in black and snow-glaring starched white, the women in a profusion of floral colors, defying the northern winter outside. Jewels sparkled, every movement setting off a brilliant flare.

The king, Gustavus VI, gave the presentations, scrolls and the Nobel medal. Each man bowed to the king and returned to his seat. When Shockley came up, Emmy, wrapped in a fur stole, beamed with pride. May's eyes never left her son.

When all had been honored, the orchestra played the Swedish national anthem and the royal family departed.

Figure 18 Shockley receiving the Nobel Prize from King Gustavus VI.

Shockley met Emmy and May in the auditorium anteroom. He gave up his scroll and medal to an official (who delivered it to the City Hall where they would all be on display) and returned to the hotel to freshen up. On the way out, May missed a step and sprained her ankle. Years later, Shockley told his secretary she did it on purpose to attract attention.[124]

Because of world conditions, the Swedes decided to reduce the pageantry. The official banquet usually was held at the city hall (Stockholms Stadshus) with about 1,300 guests, 250 of them students. That year, the dinner party was only 175 and was held at the stock exchange. The guests and new first Laureates assembled in the library where they were individually introduced to the king and queen. A line was formed, each lady with a man on her arm to escort her into the dining room.

The King of Sweden offered his arm to Emmy Shockley, and the two walked in at the head of the line.

Shockley escorted the wife of a Nobel official. Jane Bardeen found herself escorted by the Duke of Södermanland, and Keren Brattain by the Norwegian ambassador to Sweden. They walked to a long table. The ladies were seated, and then the men took their seats, everyone behind their name tags. Brattain's son Bill discovered one of Cournand's daughters, Marie Claire, nearby and was distracted for the evening.

Gustavus offered the first toast, then each Laureate, in reverse order to the award order, made a toast. By the time they got to Shockley, he suggested that everyone had said everything there was to say, and very well too, and he hadn't a thing to add. With the aquavit toasts, the party was soon mellowed out (except presumably for Semenov's thugs, and the king, who drank only water). The menu, as listed by Brattain, was Niersteiner Spätlese 1949 with salmon, Chateau D'Angelus 1953 with turkey, and then Lanson Pere & Fils Brute, Carte Noire ('champagne, no less,' Brattain reported) with dessert. Then came coffee and cognac, and everyone was mellower still.

The student procession began after dinner, the young people singing beautifully. A girl made a welcoming speech in English; Sir Cyril responded with thanks, and the students left. Everyone hit the bar for a nightcap and then returned to the hotel. Shockley and Emmy came in last. They had been left behind in the shuffle. They all consumed more champagne at the hotel and 'we certainly were in a hilarious frame of mind,' Brattain admitted. The new Nobel Laureates finally went to bed, but only after the hotel management switched the lights on and off at 2 a.m.

The men assembled the next morning at the Nobel Foundation office to pick up their award checks, $12,083.35 each for the three physicists.

The day after the awards, Laureates traditionally give the Nobel lectures. With few exceptions these lectures, which the Swedes meticulously record, are justifiably forgotten. Essentially, the Laureates describe what they did to win such an honor, which of course, everyone already knows. The physics speeches were given at the university before the Royal Academy of Sciences.

Bardeen went first, then Shockley, followed by Brattain. Shockley spent most of his 45 minutes thanking everyone for making the event even more memorable than he dreamed of, spending so much time with gratitude he had to hasten through the scientific part and ran five minutes over, an atypical response by the usually impassive William Shockley. May, now on crutches, stayed behind and missed her son's speech.

The Laureates rushed back to their hotel to change for the palace dinner. The women had brought along two dresses as it was considered bad manners to appear before the king twice in the same dress. The men could make do with the standard formal attire.

The order of processional was changed. Shockley escorted the queen this time and Gustavus brought in Mrs Cournand. Bardeen got a princess, Sibylla, the king's daughter-in-law and mother of the crown prince. Only wives were allowed at this function, so May stayed back at the hotel again, undoubtedly pouting.

Dinner was lobster and reindeer. The gentlemen retired afterwards to the king's private apartment for cigars and cognac. Gustavus, the last Swedish king with any political power, was an amateur archeologist and an expert on Chinese art, so the royal apartments were cluttered with stuff. While he was proud of his collection, the queen lamented how much trouble it was keeping some semblance of order, although it's hardly likely she did the cleaning.

The next evening was left to the Nobel committee for a dinner and for the first time the men could get out of their formal attire for regular business suits. Shockley, sitting next to the hostess in the place of honor, gave the after-dinner toast. He was told to do it before dessert. The men had been coached on the serious Swedish rules of who 'skoals' whom and when (one is not permitted to skoal one's hostess, for instance, since it is her duty to skoal everybody else, and if any other person was allowed to get in a toast she could soon be too tanked to fulfill her duty).

Shockley and Emmy left their hotel room doors unbolted that night as instructed. At 7:30 in the morning, the door opened and two beautiful girls came in: one, carrying a silver tray of coffee and cookies, wore a crown of burning candles; the other carried a single candle. They stood at the foot of the bed and sang a hymn. It was the festival of St Lucia.

That night Sweden's university students took over the entertainment. Their banquet was run by the student's Society of the Ever-Smiling and Leaping Frog and festivities began when the society's namesake, a cigar dangling from his mouth and one eye blinking, entered the dining hall. The students sang the frog's anthem and placed him on the stage, whereupon the skoaling began anew.

Shockley sat next to the student secretary of the organization. The second toast would involve everyone standing on his or her chair. But Shockley got up on his chair first, demanding attention by clinking a spoon against a glass. He proceeded to pluck a flower out of the air and presented it to the hostess. He did it again just to show he could.

After that toast, a cluster of girls in their white Lucia costumes entered the room and sang to the Laureates. The students performed a skit in Swedish (with English synopsis provided), which included the squeal of a transistor oscillator, a calculating machine and atomic energy, all in a medieval setting. The Laureates were presented with official scrolls from the society and finally, exhausted at midnight, left for their hotels.[123]

Nobel Week finally was over.

The Shockleys went to Göteborg for a day so Bill could give a lecture. May, still on crutches, caught a serious cold bordering on pneumonia, probably because few of the buildings in Göteborg were well heated. They flew to Germany where Shockley had other lectures to give, dropping May off in Wiesbaden with one of Lou Vee's children.

The Shockleys returned to California before Christmas.

Shockley was now a Nobel Laureate and as famous as a physicist could be. He had achieved the pinnacle of a scientific career, and if there were people out there who did not like him, or thought his award undeserved, he had the Nobel Prize and they likely didn't. No matter what a Nobel Laureate did for the rest of his or her life, the title remained. Bill Shockley was probably as happy then as he ever would be again in his life.

In classic Greek tragedy, the stages of the hero's life are three: *moira*, named after one of the Fates who controls our destiny; *hubris*, the pride that precedes the fall; and *nemesis*, for the god of retribution, who demands payment for *hubris*.

For Bill Shockley, *nemesis* came with quick, wrathful strides and venomous fangs.

PART III
Nemesis: Silicon Valley and obsession

CHAPTER 9
'Really peculiar ideas about how to motivate people'

Jim Gibbons walked into Shockley's office, sat across from him and was ready when Shockley pulled out a stopwatch.

'You have 127 players in a singles tennis elimination match,' Shockley said. 'Obviously, you've got 63 matches and only 126 players can be in the first round so there's a bye. You can put that next guy in so you have 64 people in the next round and you have 64 matches. How many matches does it take to determine a winner?'

Click.

It was August 1957. Jim Gibbons, a young physicist, like every other new employee had to take a little intelligence test. Shockley knew perfectly well that Gibbons had a PhD from Stanford, worked at Bell Labs and won a Fulbright scholarship to Cambridge University – a good sign he had something between his ears besides lint. But everyone coming to work for Shockley Semiconductor Co., had to take a battery of tests, either with Shockley in Mountain View or with a New York testing agency. Shockley had great faith in this kind of testing, feeling increasingly that things like intelligence and creativity can be quantified. He had begun exploring their uses while still at Bell Labs and became a firm believer. That the tests had no real scientific basis never seemed to bother him.

Gibbons thought only a few seconds and said, 'Well, it must be 126.'

Click. Shockley looked down at his stopwatch, his face reddening.

'What?'

'Well, it must be 126.'

'How did you do that?' Shockley asked, his agitation growing.

'There's only one winner and that means 126 people have to be eliminated. It takes a match to eliminate somebody, so there must be 126 matches,' said Gibbons assuredly.

Shockley pounded the table in fury.

'That's how I'd do it! Have you heard this problem before?' he demanded.

'No sir,' said the young scientist, confounded at Shockley's reaction. The Nobel Laureate was coming unhinged.

Shockley gave him another problem, again clicking the stopwatch into action. Gibbons thought about this one but could not figure out a quick answer. As time elapsed and he said nothing, Shockley's face returned to its normal color and he sat back. 'You could feel the tension start relaxing,' Gibbons remembered later.

'That's enough, Jim. You're now at twice the average time for the lab to solve this problem. Let me tell you how you do it,' Shockley said, his equilibrium restored. Gibbons had missed the key.

'It was really tough for him, the fact I got the first one,' Gibbons says. 'He'd set the damned thing up so you'd say $63 + 32 + 19$ – you just don't sit there and say "126."'

The possibility that this young man – Gibbons was in his early 20s – could compete clearly upset him. Gibbons only redeemed himself by failing the second test. The thought that Gibbons might have been as smart as he was ('Not even remotely close to being true,' Gibbons said), seemed to frighten him. 'If I'd seen the next trick, the guy would have been apoplectic,' Gibbons remembered.

Gibbons did well enough with the rest of the test and walked out of Shockley's office to the laughter of the other researchers in the building, all of whom had faced the same test.[1]

The rise and fall of Bill Shockley's company took less than a year-and-a-half. It profoundly affected Shockley, but had even more impact on the world around him and on our lives today. In all of the history of business, the failure of Shockley Semiconductor is in a class by itself.

Shockley picked his backer well: he found an honorable man. Arnold Orville Beckman was born the last year of the 19th century in rural Illinois, somewhere between Kankakee and Bloomington. He played the piano in the local nickelodeons while in grammar school and by the time he graduated from high school as valedictorian, he already was making more money than his father. He got chemical engineering degrees from the University of Illinois and then went to Caltech for his PhD. He stayed and taught there. Beckman acted as a consultant in his spare time, working with the National Postal Meter Company in devising special ink, and then forming his own ink company. He owned 10% of his National Ink Appliance Company; National Postal owned the rest.

By 1940, National Ink was doing so well that Beckman quit his teaching job and changed the name of the company to Beckman Industries. In 1952, the company went public. The investment business of Lehman Brothers handled the deal, but only on condition that Beckman remain at the helm. When all was done, Beckman himself had 40% of the company. By 1953, Beckman Industries was earning $21 million, making everything from guided missiles to seismographs.[2]

The business alliance was a natural: Shockley and Beckman had strong Caltech connections through which they apparently got together. Beckman was particularly interested in automation and he believed the secret to it was the transistor.[3] Shockley's reputation as a physicist was unequalled, his knowledge of semiconductors unchallenged, and he was anxious to get rich in business. His reputation as a manager, unfortunately for Beckman, had not caught up with him, and if Beckman checked with Bell Labs, he asked the wrong people.

Beckman preferred that Shockley run the transistor business in Southern California, probably somewhere near Caltech, where it would be close to the rest of his company. But Shockley was sold on Northern California. He grew up there, his mother was there, and he had friends there as well. Also important to him, probably, the region was and is – not counting an occasional earthquake or two – an exquisite place to live.

The apricot and prune orchards of the Santa Clara Valley had not yet been flattened for developments and strip malls; the air was clean and fragrant, and smog was a thing of the future. It remained largely Jack London's 'Valley Of Heart's Delight,' with hills and fields sere and flaxen in the summer and fall, vivid green, almost chartreuse in the winter and spring.

It now is widely recognized and well-documented that a reason the Santa Clara Valley, which surrounds Palo Alto, has been able to survive challenges to its technological and economic hegemony from other areas in the country is largely a matter of climate and geography. Engineers and physicists, who could work wherever they chose, overwhelmingly preferred then (and prefer now) to take jobs there. Shockley may have been one of the first to understand that, particularly with his love of the outdoors.

Forty minutes to the north is San Francisco, one of the world's most beautiful cities, in the mid-1950s about to enjoy yet another golden period. It had first-rate symphony, opera, theater, and the beginnings of the coffeehouse culture. Kerouac and Ginsberg read poetry while the Kingston Trio played ersatz folk music at the Hungry i.[3,4]

Even more enticing, perhaps, was the lure of Stanford. The key figures here were Frederick Terman, provost of the university and professor of engineering, and John Linvill, a young engineering professor. Terman was the son of Lewis Terman, the Stanford psychologist whose 'genius' test Shockley failed as a child.

Linvill had been an assistant professor of engineering at MIT in the early 1950s, and had taken a year's leave to work at Bell Labs, mostly to study transistors. He spent little time with Shockley there but knew of him. 'Shockley at Bell Labs was just as visible as hell.' Linvill attended several seminars that Shockley ran, and read the internal research papers out of Shockley's group. He went to Stanford a year later. When he learned of Shockley's intention to start his own company, he immediately thought of an alliance and enlisted Terman. Terman suggested to Shockley that a relationship between Stanford and the new firm might be mutually useful.[5] Linvill even checked the real estate market for him and came up with three places that might suit the new company.[6]

Terman firmly believed Shockley was seeing the future and wanted Stanford to be involved. He particularly wanted Stanford students and professors to understand semiconductors and semiconductor manufacturing, and no one in the world knew as much about this as did Bill Shockley.[5] Terman already had plans to put one of his best graduate students in Shockley Semiconductor to act as a conduit to transfer the technology to the school: Jim Gibbons, then just finishing his PhD.

Terman was hatching his own glorious scheme, the first university-owned industrial park in the world. Stanford's founding father, Senator Leland Stanford, had left the school a huge amount of land, only a small portion of which was used for academic purposes.* Much of it was

---

* The school was built as a memorial for the Stanfords' 15-year-old son, who died of typhoid fever. Its formal name is the Leland Stanford Jr. University.

(still is) undeveloped, rolling hills covered with grassland, madroñe and stands of redwood. Terman's idea was to turn some of that into an industrial park so that the firms and university could cross-fertilize. Stanford would own the land, and the companies would erect their own buildings and lease the ground for 99 years.

Terman had willing tenants. In 1938, he put two of his graduate students, William Hewlett and David Packard, together to form a company, even providing seed money from Stanford to help them along.* After the war, they moved into the industrial park. Others with Stanford connections also were willing to rent space. The plan worked as Terman hoped, with firms hiring his students, either as interns or full-time employees, contributing to joint research projects, and donating equipment. The university provided faculty appointments for company researchers and executives and sent professors to learn business.

Stanford already was turning out physicists of considerable note, and had begun, under Terman's direction, concentrating on engineering. Between 1950 and 1954, Stanford awarded 67 doctoral degrees in electrical engineering; the much-larger University of California at Berkeley, up the road, awarded 19. The new engineers would provide considerable intellectual capital for the area. Terman also used the contacts he made during the war to bring Pentagon research and procurement dollars to the area. Between 1950 and 1954, California got 14% of all the Pentagon's funding, about $13 billion. Thanks to Terman, much of it went to the Santa Clara Valley. Several companies set up shop, including Sylvania, Fairchild Camera and Instrument, General Electric, Philco-Ford and Westinghouse. Lockheed too started a huge research arm for weapons and missiles in the Stanford industrial park.[7]

Shockley's company wouldn't have to be isolated from the rest of Beckman Industries. Beckman was building a facility for his Spinco division at Page Mill and Junipero Serra Roads at the northwest corner of Terman's park. Spinco built centrifuges for hospitals and medical laboratories. When the new facility was finished, there would likely be room for Shockley Semiconductor.

---

* Hewlett-Packard's first customer was Walt Disney. They provided the equipment for the stereophonic sound for the film *Fantasia*.

Beckman acceded to Shockley's judgment. It is largely because of that decision that Silicon Valley is concentrated in the Santa Clara Valley in Northern California, not in Pasadena.

On 14 February 1956, Beckman and Shockley held a news conference in San Francisco to make the official announcement of their plans. By now several lazier newspaper reporters were calling him the inventor of the transistor, ignoring Brattain and Bardeen, although Shockley did nothing to encourage that. He predicted that the devices would replace vacuum tubes and that transistor production would increase by one hundred to one thousand-fold in the next five to ten years, which turned out to be a major understatement. The *Daily Palo Alto Times* quoted 'local electronics firm spokesmen' as giving two reasons why the Beckman–Shockley alliance was good for the local economy: further development of the transistor by the man who is credited with its invention, would 'open new horizons for the industry,' and local firms would have an edge exploiting the advances made by Shockley and his team.[9] That would be a historic understatement.

When reporters asked him why he was leaving Bell Labs, he said, 'You only live once. I would like to do something else for a change.'[9]

From the beginning, Terman and the university were deeply involved in the project and gambled considerable effort and prestige on Shockley.

Shockley also mentioned in his talk the possibility that transistors would revolutionize the use of 'electronic brains,' to the point that people might even figure their income taxes on the machines. In 1957, few people besides him would have made such a statement. Almost all computers still used vacuum tubes, and it would be another five years before transistors found their way into mainframes.

Beckman paid Western Electric $25,000 for a license on the transistor patents.[10]*

At Linvill's suggestion, Shockley rented a Quonset hut at 391 South San Antonio Road, 2,255 square feet at five-and-a-half-cents a foot. It needed work. Shockley began hiring office people to set up something resembling an organization for the scientists and engineers he was gathering, and to clean out the place. He already had three PhD physicists,

---

* The federal government had begun the first of a series of anti-trust actions against AT&T, which eventually led to the break up of the monopoly. One of the earliest results was that the company began giving away license fees.

including G. Smoot Horsley, Lee Valdes and William Happ. He had only begun the recruiting process.

Shockley's genius for selecting scientific talent was at its height. He traveled from one end of the country to the other and to Europe, placed ads in publications such as *Chemical & Engineering News*, and scoured other labs. He went to one meeting of the American Physical Society in Pasadena ostensibly to give a speech. Actually, he told the audience, he was recruiting.[11] He had arrangements with some places, such as Lawrence Livermore Laboratory, to get the names of people who had refused jobs there. That's how he found Gordon Moore.

Moore, then 28, was from the area. He grew up in the little fishing town of Pescadero on the San Mateo County coast, northwest of Palo Alto, and at the age of ten moved to Redwood City. His father was a deputy sheriff. He had a degree in chemistry from Berkeley and a Caltech PhD. With technical jobs hard to find in California, Moore found himself working on federal grants in the applied physics lab at Johns Hopkins University outside Washington, doing basic research on such things as the spectral lines of flames. He figured out the cost to the taxpayers per word of scientific articles published and concluded 'I wasn't sure society was benefiting sufficiently from what I was doing. It was time to do something a bit more practical.'[10] He put out some feelers and LLL offered him a job, but 'the work they wanted me to do wasn't that exciting,' so Moore turned them down. Shockley found his name on the LLL list and called. He told Moore he wanted to build a company that made silicon transistors.

Moore was, in old-fashioned terms, a gentleman. Tall, already balding and quiet, he was famous in his lab for choosing exactly the right field of investigation to produce the most results the fastest, and like Shockley, had an innate ability to solve problems in minutes that took everyone else months. He signed up.

So did the man who would become Moore's partner, the physicist Robert Noyce.

Noyce was handsome, athletic, gregarious and had a huge, infectious smile that lit up a room as easily as it masked a truly remarkable brain. Shockley may have been brighter; Bob Noyce was surely wiser. Profiling him for *Esquire*, Tom Wolfe wrote that Noyce had an aura about him. 'People who have it seem to know just what they are doing; they make you see their halo.' He was a walking American prototype: He grew up in a small town in Iowa, the son of a Congregational minister, a Boy Scout, a valedictorian. He went to Grinnell College, a small Congregational school between Des Moines and Cedar Rapids. Noyce got a PhD from MIT and then went to work for Philco in Philadelphia. Philco, then a major appliance manufacturer (television sets and radios), was not interested in research, Noyce felt, and he was looking around when Shockley spotted one of his publications. Shockley, like almost everyone else who met Noyce, liked him immediately. 'When he came out here to organize Shockley Labs, he whistled and I came,' Noyce said.[12] Shockley also told him he was going to build silicon transistors.

Signing on, however, wasn't as easy as just showing up. Shockley by now was becoming as obsessed with social science as he was with transistors, although he would never seem to gather the same level of critical thought. He bought into a lot of nonsense. He required everyone he employed to go to New York and take a battery of tests from the firm McMurry-Hamstra. That he required these tests is surprising, since he seemed to have the ability to instinctively recognize talent and didn't need any verification. A strange sense of insecurity seemed to be creeping in. So Moore and Noyce went off to New York, spent a day associating words and interpreting ink blots, and McMurry-Hamstra mailed the results to Shockley. Shockley made it a policy of handing everyone's results around. With all the assurance of a Ming's fortune cookie, McMurry-Hamstra told Shockley that the two future founders of one of the most successful companies in history were very bright, but they would never make very good managers.[11]

They started work in April of 1956.

Shockley hired a brilliant clutch of researchers, about a dozen bright, innovative men, exactly the kind he needed to dominate the semiconductor industry. Only one was over 30. Besides Moore and Noyce, he had Jean Hoerni, a Swiss-born physicist of prodigious talent, Gene Kleiner, whose father was one of the first venture capitalists, and Dean Knapic, who he had lured away from Western Electric to be his assistant director and production manager. He couldn't have chosen better.

Things began to all fall apart almost immediately.

Two reasons, both traceable back to Shockley, led to the problems. One was a disastrous decision.

When Shockley started, he was convinced that the future was in silicon, not germanium. Until then almost all research and devices were based on germanium with an occasional flirtation with gallium. He made it clear to all that his intention was to manufacture silicon transistors. He was not the only one to feel that way about silicon, but he was the most prestigious. Since he already had the first and most famous semiconductor research and manufacturing company, everyone who had been working with germanium stopped and switched to silicon, and everyone who had made the same decision as Shockley was assured of their wisdom. Everyone he hired, including Noyce, agreed with that decision. 'Shockley put the silicon in Silicon Valley,' says Moore.[11] Indeed, without his decision, we would speak of Germanium Valley and while the revolution still would have come, it might have come more slowly and in slightly different ways.

Then, for reasons he never explained, Shockley changed his mind. He was still going to build silicon devices, but they would not be transistors. Shockley was determined to build a four-layer diode, a device he worked on at Bell Labs.

The so-called Shockley Diode took advantage of several new concepts of how electrons could be channeled and amplified. It was ideal for certain uses, most of which at the time centered on switching. It was a p-n-p-n junction diode in four layers, with the action in the middle layers. The ideal customer was Western Electric, which manufactured the equipment for the Bell System.

AT&T operated a regulated monopoly at the time. They could design and build their own equipment and it would be as big as it turned out to be, cost whatever it cost; it did not need elegance or simplicity. The only criterion was that it worked, and since they had no competition they could control the market for the equipment. AT&T never did anything on the cheap. The company was very conservative in its engineering because it could afford to be. Shockley had spent most of his professional life at Bell Labs, working in that atmosphere. The Shockley Diode would have been perfect for the switching devices in the Bell System, he knew, and perhaps for the Pentagon. But unless he could prove they were robust to very high standards, even the Bell System wouldn't buy them. They were of very little use to anyone else at the time.

Noyce believed that starting a company with the Shockley Diode as the primary product was a mistake. The market was limited, and Shockley had no idea how to manufacture them reliably. A new company ought to make as its first product something with a broad range of customers and uses, and something within its competence. A new company could learn the culture of manufacturing that way. It also would generate sufficient cash flow to keep it healthy. Then, if it was determined useful, Shockley Semiconductor could invest in the Shockley Diode. The entire senior research staff agreed.[1,10,11]

Shockley would not budge. He ordered his staff to get to work designing and building the diodes. Noyce argued against the decision, and tensions grew quickly. They did not ease when the company found it couldn't build diodes satisfactorily. The devices had to be doped on both sides of silicon, meaning they had to be paper-thin, making them extremely brittle. The first ones off the line – those that didn't break – were unreliable and useless. Changing course at the direction of his employees was not how Shockley envisioned running a company.

By this time, Shockley considered himself expert at managing creative institutions and creative people. He had spent much of the last five years researching such places as Bell Labs, Los Alamos, and Brookhaven National Laboratory for the Pentagon, to learn how you nurture and encourage the best people. He wrote several papers on the subject. He brought his talent for operations research to bear in dissecting creativity.[13]

Shockley traced the relationship between published articles and scientific creativity and – admitting obvious exceptions – found one. 'For statistical purposes,' he wrote, 'the average number of papers per year published in recognized scientific journals is a valid index of individual productivity.' He pointed out this was even true of most of the most famous scientists. Louis Pasteur published 172 papers in his professional lifetime, Michael Faraday 161, Louis Agassiz 153. Moreover, the variation in productivity could be as much as a hundredfold between the best people and the mediocre. Most of life isn't like that: variations between individuals are much less dramatic. The difference between a poor hitter in baseball (one who hits twice for every ten times at bat, or a .200 batting average), and a legendary one (a hit four out of ten times, or a .400 hitter) is two to one, and there hasn't been a hitter that good in 65 years. In creativity, the ratio, Shockley found, was a hundred to one. The more ideas a person can handle, the likelier he or she would be to invent something useful.[14]

He criticized the government for not paying its scientists well enough, either, particularly its managers.[17] He was determined he would pay his employees well and he did. He considered himself their benefactor.

Shockley reduced the process to logarithms and charts. He was working on several of these papers while his company was falling apart. All his charts and logarithms, the 'mental temperature,' did not tell him how to treat or lead the brilliant people he hired.

Shockley firmly believed that scientific advancement was the result of a solitary genius or at most a small group of geniuses who set the stage for an intelligent team of researchers below them to break the necessary ground, a kind of trickle-down creativity. The coterie of great minds running the Manhattan Project stimulated the worker ants below them to great achievement. He gave little credit to creativity from below.

Shockley had a model of how laboratories and institutions should work that very clearly involved that kind of hierarchy: The workers take direction from above (their betters?) and progress ensues. He, of all people, should have known better. The invention of the point-contact transistor violated that model (the men responsible worked for him without much direction), and if he had paid attention to history he would have seen that most innovation comes from motivated individuals, not teams or hierarchical dictates.

To be fair, no one else knew how to handle such a group either. Traditionally companies worked from top to bottom. Only when innovation became coin of the realm did different models become imperative. In the 1970s and 1980s, IBM described its philosophy for managing creative people as 'wild geese flying in formation.' They were free, but they had to move with the other geese in the direction the head goose pointed. Shockley wouldn't have put it that way, but that was his model too. It didn't work for him then; it didn't work for IBM later.

In truth, he had no idea how to manage.

'He had some really peculiar ideas about how to motivate people,' said Moore. 'First of all, he was extremely competitive and controversial. If there were two ways of stating things, one of which was controversial and one of which was straightforward, he'd pick the controversial one every time. He just thrived on stimulating controversy.'[10] That stimulated conflict, not originality.

The problem certainly wasn't his scientific leadership. Gibbons remembered sessions at Kirk's restaurant, a Palo Alto institution famous for huge, cheap hamburgers, with Shockley scrawling formulas and notes on napkins splattered with hamburger grease and wishing he could save some of them. 'He did have this way of taking a problem that's eight-feet thick on this side and everybody's going whack, whack, whack, and he turns that thing around and finds a place where you can go and it falls apart,' said Gibbons. 'He was uncanny in his ability to do that. I've worked with a lot of Nobel Prize winners by now and a lot of pretty smart people. I've never seen anybody that could do it this way.' One physicist swore Shockley could actually see electrons – he knew too much about them.[1]

He had trouble seeing people. Whether he always had this flaw or this was something that happened after the phone call from Bardeen and Brattain is unclear. It did not appear to have been a problem during the war.

'He was very attractive to bright young people,' Terman later explained, 'but was hard as hell to work for.'[12]

One man, R. V. Jones, resigned weeks after being hired.

Shockley quickly began feuding with Dean Knapic, calling him a 'pathological liar.' Beckman was so disturbed at the quarrel he asked the New York psychological testing company, McMurry-Hamstra, for help. One partner wrote back that going over Knapic's original tests, they did not agree with Shockley's assessment that Knapic lied. On the other hand, 'there is considerable cause for concern in this record, and I would suggest that it be subjected to such further investigations as may be necessary to remove or substantiate those doubts.' After investigating Knapic's background, and finding no conclusive proof that Knapic lied, Shockley retreated, but the animosity between the two men lingered and Knapic held a grudge.

Shockley was often insulting, treating his employees the way he treated his sons, with no glimmer of sensitivity. His favorite crack, when he thought someone wrong, was: 'Are you sure you have a PhD?'

A kind of insecurity had crept into a man formerly so sure of his own intellectual prowess. 'The relationships were not good,' Noyce recalled. 'I think the main problem I had with Shockley was that if you had done a piece of work, then he would call up Bell Labs and check on it to see if that was correct or not. He didn't trust you, was the way we interpreted it. He was insecure enough himself so that he had to rely on other authority rather than his own resources. On the other hand, scientists like to check their ideas with somebody else to see if they will fly, so that could have been a more charitable interpretation.'[12]

Charity became hard to find.

He screamed insults at one researcher, Jay Last, with many of Last's colleagues within hearing range. Metallurgist Sheldon Roberts threatened to quit. So did Hoerni. By February 1957, less than five months after the company was formed, the dissension made its way into Shockley's note pads. 'Felt it would be catastrophic if CSR & JH left,' Shockley noted.[15]

Worst of all, he could not keep himself from believing he was in a competition. Just as he had set himself up against Brattain and Bardeen at Bell Labs, he now exhibited the same behavior against his own employees, the very people he hired because they were so bright. He just didn't want them to be as bright as he was.

'He asked the group one day what he could do to make their jobs more interesting and more rewarding,' Moore recalled, 'and a couple of them commented, "gee, we'd like to be able to publish some papers." So he said, "okay." That night he went home and he worked out some theory for the effect on semiconductors, came back and handed it to them the next morning and said "here, flesh this out and publish it." Typical of the feeling he had for what it was that motivated people.'[10] That his employees could come up with their own ideas did not register with him.

Once Gibbons and Shockley were working on a paper together. Gibbons had developed a new, elegant way of designing an avalanche transistor, which would have been helpful in the Shockley Diode project. Shockley left for Europe on business and told Gibbons he would read the draft on the plane. He sent back some comments. 'I thought, well, that's interesting,' Gibbons said of the comments, 'it's a different way to look at it, not a lot better. I didn't think it was any better. It was an embellishment that detracted from the real core.' When Shockley returned, he asked how the paper was doing. Gibbons said he hadn't made any of the changes; he was waiting until Shockley got back.

'I wanted to know why you thought this the right thing to do based on what I thought was my model,' Gibbons explained.

'It's not your model,' Shockley snapped. 'If you are not smart enough to see the improvements I wonder if you're smart enough to be working for me!'[1]

Gibbons said that kind of insight that happened with the avalanche transistor didn't happen to him often, but when it did, Shockley admired it. Then 'he always had to prove that I didn't have it quite right.'

Once Shockley decided to exert the main effort toward a new device, he determined for inexplicable reasons that it would be a secret within the company. He said it was as important as the invention of the transistor, which it wasn't. Only those people directly involved could know what he was building. That included Noyce, but didn't include Moore. The problem was that Shockley Semiconductor was in a fairly small building and the laboratory essentially consisted of one small room. No one could truly keep a secret under those circumstances,. The only way the scientists could follow Shockley's orders would involve offending – unnecessarily – those sitting next to them who were not involved.[10,11] The device never panned out.

One day a secretary cut herself on a pin stuck in a door. Shockley was sure someone had done it deliberately. He had been getting late night telephone calls where the caller hung up when he answered, and his feeling of threat was not entirely unjustified. He was sure that someone had left the pin in the door to harm someone, and he thought it was one of two technicians. Shockley ordered every employee up to San Francisco to take a polygraph test. The first man went and came back 'exonerated.' Every other employee flatly refused to go. Shockley, faced with complete defiance, had to back off. It turned out the pin was the remains of a thumbtack. Someone had posted a notice on the door and the head of the tack fell off, accidentally leaving a sharp end.[10,11]

Meanwhile, Shockley Semiconductor wasn't doing very well. It had no product to sell and had virtually no income or customers. Western Electric would have been interested in the Shockley Diode if the company could produce sufficient numbers of reliable devices; ditto the Army Signal Corps. But so far the company couldn't do it. Noyce was still insisting the diode was the wrong technology and he and several others took time off to work on the silicon transistor, which probably didn't help productivity.

The chronology of the great mutiny is unclear. The only notes are those Shockley wrote in his 'Golden West' brand notebooks, an informal diary. The rebels left none. But apparently, this is what happened.

By May 1957, Beckman was having some of his own problems. His company was in an industry already famous for cycles and his company now was in a down cycle. The amount of revenue being spent on research and development was at 8%, and earnings were down. So was his stock. One place spending the most on research, proportionately, was Shockley Semiconductor, for small things such as typewriters and paper. He flew up to Palo Alto and called a meeting at the company with Shockley and his senior people in the room. He explained to them they needed to keep some eye on expenses.

Bill Shockley responded with profound stupidity.

He stood up and told Beckman, in front of the other staffers, that he found what Beckman had said outrageous and unacceptable. 'Arnold,' he said, ' If you don't like what we're doing up here I can take this group and get support any place else.'

Then he walked from the room, leaving behind an astonished senior staff and a humiliated benefactor. Beckman extricated himself from the room as politely as possible and flew back to Los Angeles. The senior staff gathered in clusters to talk about the amazing confrontation.

The next morning, eight or nine of the staff decided it was time to let Beckman know what was really happening at the company. Moore was elected. He went to one of the others' houses and called Beckman.

'That's not a serious threat,' he told Beckman. 'Shockley couldn't take the group with him if he wants to at this stage of the game.'

'Things aren't going well up there, are they?' Beckman asked.

'No, they really are not.'

Beckman volunteered to fly back north to meet with the discontented, without Shockley's knowledge. Eight men came to dinner with him on 29 May. They met three or four more times – no one remembers exactly.

The message from the researchers was simple: Shockley had to go. They were prepared to quit en masse otherwise. That would have left Beckman's subsidiary essentially without a senior research staff. They were amenable to a compromise, however, and over the course of the meetings they worked out one satisfactory to them. Beckman would use his influence to get Shockley a permanent teaching position at a university, probably Stanford; Shockley would remain as a senior consultant but not as director; Beckman would send up a professional from his company to manage Shockley Semiconductor; and the scientific and technical decisions would be made by a committee of researchers, headed largely by Noyce. Everyone left the meetings with the feeling that Beckman had agreed to the compromise. He even suggested one of his people, Joe Lewis ('I'll never forget that name,' Moore said), to manage the company, and the rebels thought he sounded perfect.[10,11] Lewis even came and visited again without telling Shockley.

Beckman probably felt badly that all this was happening behind Shockley's back, and decided it was time to tell him. Around the first of June, Shockley got a call in his office from Beckman asking that he and Emmy join him for dinner at the Jack Tar Hotel in San Francisco.

Shockley, sure it was purely social, called Emmy and told her happily to dress up.

After polite greetings, they all ordered drinks to wait for a table in the dining room. Beckman, as usual, was direct. 'I have bad news,' he told Shockley. 'I'm sorry to have to tell you this.' He then outlined what had been happening: that most of his PhDs were on the verge of leaving and that Shockley's management was the issue. He implied that the decision was up to Shockley: if he were to stay, they would go; if he went, they would stay.

Shockley was stunned. Either his competitive or paranoid antennae had missed the signals or he was in serious denial – Emmy's strong impression was that all this was unexpected. He said almost nothing.

The waiter came to tell them that a table was ready and they went to sit and order dinner. Emmy picked at her food while Shockley and Beckman talked. Beckman told Shockley of the proposed compromise, essentially giving Shockley a way out of his dilemma. The compromise meant that Shockley would end up a consultant to his own laboratory.

Beckman kept urging Emmy to eat, but heartsick, she could not. She asked her husband if he could start again if the eight left. He said he thought he could. He was sure Smoot Horsley was loyal and would stay, and perhaps a few of the others, but he was going to lose every PhD except Horsley, he feared.

'Are you prepared to help Bill with this?' Emmy asked Beckman.

He evaded the question.

'Well, maybe we can start again somehow,' she said. Her role, she decided on the spot, was to help Shockley do whatever he wanted to do. As simple as that, she said later.[16,17] She never deviated from that policy.

They drove home in silence. The next morning, Shockley went to the lab to confront the rebels. His intention was to call in each singly to find out who was in the rebellion and whether, perhaps, Beckman was exaggerating the problem. Moore said he thought Shockley called them in the order he thought loyal to him, apparently still not understanding the depth of the problem. The first one in was C. T. Sah, an engineer. Sah told him truthfully that he was not involved in the rebellion and knew nothing about the dinners with Beckman. The next one invited to Shockley's office was Gordon Moore.

'I had the privilege of informing him that, yeah, I was part of the group and so essentially was everyone else on his senior staff.' There was no use going through the rest of the group, he told Shockley. They were all in on it.

Shockley got up and left the office.

That night, at about seven, he walked in the door of his home. Because of he look on his face, Emmy knew the answer to her question before she even asked it.

'Was it true?'

'Yes.'

He went to a settee in the living room and lay down. The two-person couch was too small for him, so his legs hung over one of the arm rests. Never in her life, even as a nurse, had Emmy seen anyone's face as white as her husband's.[16,17]

Even Shockley knew he had failed.

That Shockley was surprised is itself surprising. The signs of trouble littered his diaries. Moore warned him of mental stagnation among the researchers. Noyce told him they were spending time on the wrong product. Yet he did not take any of it seriously.

Beckman had a surprise for the rebels. Within a day or two of his dinner with Shockley, he called another meeting. This time he told the eight he had decided Shockley would remain in control, but he hoped something could be worked out. The rebels, who thought they had a deal, now found they did not after all.

They were unprepared for the change in position. Moore said later he had heard gossip that Beckman talked to someone at Bell Labs and had been advised that if Shockley were dumped from his company it would destroy his career. That seems unlikely. Few at Bell Labs cared that much for Shockley to intervene, except perhaps Kelly (who was only partly sorry Shockley left) and Fisk. Plus, his career in the heady world of research physics would probably not have been affected – his reputation as a scientist was as yet undiminished. Business and physics then, however, were two different worlds. Whatever the reason, the rebels now were in a very tenuous position.

On 3 June Noyce asked to speak to Shockley. The discussion was very 'factual,' Shockley wrote in his notes, but clearly he and Noyce disagreed on major policy issues. Noyce told Shockley the reason they went behind his back was that they felt they couldn't talk to him. They were not 'out to get him,' Noyce said, and they didn't want to be fired. 'I believe it can still be made to go successfully. I prefer to try to do so,' Shockley wrote in his diary.[18]

On 6 June Beckman called Shockley.

'He was happy and relieved to know this situation had hit hard,' Shockley wrote in his Golden West notebook. 'He feels W=S and AOB have had a good relationship.' (He had begun referring to himself in his journal in the third person.) Beckman asked Shockley what he thought of the research group. He told Beckman that he thought Noyce was good technically 'but immature.' Shockley thought Horsley was the better physicist. The rebels, he said, were 'immature, not aggressive leaders, look[ing] for leader in vacuum.'

Shockley had prepared for the conversation, writing notes to himself beforehand. He wrote that he had not sensed the 'seriousness of the situation,' being more focused on production.[18] 'It may work out,' Shockley wrote in his diary. 'I hope it does. I shall not be sure. This is the way I am. This is one of my scientific assets,' he wrote to himself.

The new plan was worked out with Beckman:

- No one was going to get fired for at least six months.
- The administration would be structured so that the scientists had a larger say in what was happening, with an interim committee making decisions until Beckman could send up a professional manager to take over.
- Non-technical decisions would be in the hands of the manager.
- Beckman would, for the first time, assert final authority over what happened at the company.
- Shockley would get a new contract with new definitions by 3 September.

Beckman sent an aide, Maurice Hanifan, the manager at Spinco, to act as his liaison with the company. An honest man in an impossible situation.

In August, Shockley, Emmy and his two sons went to Woods Hole, Massachusetts, for a physics seminar. Shockley was not without loyal spies back in the shop. One, Elmer Brown, wrote a letter telling Shockley that the compromise wasn't working. He detailed a staff meeting with Hanifan. Knapic (not one of the eight) was using an organization chart from Hanifan, who said he didn't believe in such things. Noyce, reported Brown, was still bristling that Shockley was making technical decisions, still convinced Shockley was chasing the wrong technology. Hanifan reminded Noyce that Shockley was still director and was going to continue in that capacity, Brown wrote.[23]

The move to the Stanford Industrial Park began when Horsley led the first contingent out of San Antonio Road to the Spinco plant. Soon they were actually turning out usable diodes: 72 the first week, 200 the next. They wanted to be able to increase production to a thousand a week. But back at the old shop Noyce and colleagues were working on manufacturing processes for silicon transistors, probably in defiance of Shockley's orders. Essentially, Shockley Semiconductor had split in two with no one in charge.

The situation was untenable for the rebels at Shockley Semiconductor.

The Shockley family had a propensity for noting catastrophic events in as few words as possible, as in May's '8:20 Wm. died.' In September 1957, her son made the following entry in his notebook: 'Wed 18 Sept – Group resigns.'

It was the birth notice of Silicon Valley.

What happened to the eight is not a digression in the story of Bill Shockley. It is the key to understanding the rest of his life. They became known in the mythology of the valley as the 'Traitorous Eight.' Emmy denies Shockley ever called them that, and no reference to his having done so has been found. But they jokingly wore the label themselves after a while.

When they walked out, the men already had a fallback. They had decided they liked working together and believed that Noyce had a rational handle on how a semiconductor company ought to act at this stage of the industry. He also had some interesting ideas for pushing the technology. Gene Kleiner's father had connections at the New York firm of Hayden Stone, and Hayden Stone agreed to act as facilitators for their plan. Two men, Bud Coyle and a young Harvard MBA, Arthur Rock, flew to California to meet them. After a long rambling bull session, Rock and Coyle convinced them that instead of going to work for another company they ought to form their own, perhaps finding another firm to set them up as an independent subsidiary – as Beckman had done for Shockley.

'We found that fairly easy to accept as an idea because we all lived here, we all owned houses in the area, it would clearly be a lot less disruptive to our personal lives than any other solution,' Moore said. The goal was to find ways of pursuing Shockley's original goals.

The men sat down with a copy of the *Wall Street Journal* to identify companies that might be willing. They went through every name on the Stock Exchange, and came up with a list of 35 possibilities. Hayden Stone representatives visited every one. All 35 turned them down.

Then Hayden Stone found Sherman Fairchild.[10,11] He was perfect for them.

A tinkerer and great lover of gadgets, Fairchild earned more than two dozen patents in his career without the benefit of a degree in either engineering or science. He invented the first aerial camera when he was 23, and even designed a plane to carry it. He was one of the largest stockholders in IBM and his father was one of the largest stockholders in Pan American World Airways. Fairchild designed and sold tape recording systems and even invented a match that would not blow out in the wind. He did it for the love of puttering, not, as he often averred, to make money. He made money nevertheless, a fortune of more than $80 million. He also loved jazz (he played a hot piano), architecture (he designed his own home) and good food (he studied at the Cordon Bleu). A bachelor, he was not averse to having a beautiful woman share his activities.[20] Life was a great adventure for Sherman Fairchild

Coming up with $1.5 million for the rebellious eight, a risk that scared off everyone else, seemed perfectly normal. The deal was that if, after two years, the company failed, Fairchild would lose his money. If it was successful, he had the right to buy them out for $300,000 each. This was 1957, and $300,000 was a lot of money. The day after they left Shockley Semiconductor, they signed the contract. Within a month, they had a facility on South Charleston Road, a continuation of South San Antonio Road – down the street from Shockley.

On 4 October the Soviet Union launched Sputnik 1, the world's first space satellite, which terrified Americans and convinced them they no longer had a technological lead over the rest of the world, especially the evil Communist empire. Sputnik 1 (and Sputnik 2, which carried a live dog into space) turned America's attention toward science and technology in a way that decades of nagging could not. More importantly, Congress, then led by Senator Lyndon B. Johnson of Texas, insisted that the conquest of space should be a major government priority as a matter of national security. That meant that a lot of transistors were going to get sold. Shockley and his company were just as well positioned as Noyce and Moore at Fairchild, but Noyce was selling the right product; Shockley wasn't.

Noyce, 29, was head of research at Fairchild Semiconductor; Moore was head of production. By the fall of 1958, a year after they split from Shockley, they were selling transistors to IBM for $150 a piece, packaged in empty Brillo cartons. The devices were manufactured using a form of etching, photolithography, set up by Noyce, the first commercial use of the technology. By December, a year and three months from the split, they had earnings of better than a half-million dollars and were making a profit. Shockley Semiconductor still had no reliable product and was still not earning money.

In January 1959, Noyce came up with the notion of protecting the junctions of a transistor under a coating of silicon dioxide, using the breakthrough 'planar' manufacturing process developed by his Fairchild co-founder Jean Hoerni.* By 30 June they could patent the first integrated circuit. They knew something about this idea because the silicon dioxide layer had been developed at Bell Labs and Shockley sent them a memo about it when they worked for him.

Fairchild, knowing a winner when he had one, exercised his option and bought the company. As the eight rebels found themselves financially enabled for life, Fairchild Semiconductor fell apart. The orders started coming from New York, from people who had no idea what the industry or technology were about. Although Noyce was promoted, he found himself increasingly cut off from the decision-making. Several of the original eight left either to form their own companies or to start a new adventure. They became known as Fairchildren.

In 1967, Fairchild had its own rebellion, led by its manufacturing director, Charlie Sporck, who set up his own company at National

---

* Jack Kilby at Texas Instruments patented a similar device earlier in the year. Noyce's differed in that it used a planar process developed by Hoerni, which hid all the contacts within a cocoon of SiO_2. Kilby and Noyce are considered the independent co-inventors of the integrated circuit. In some ways, the difference between the two devices is similar to the differences between the point contact transistor and the junction transistor. The Noyce device eventually became the standard. In 1955, the two firms agreed to cross-license their patents and Kilby eventually won a Nobel Prize. Noyce died in 1990, before he could win his own prize. He was 62.

Semiconductor, spiriting away several Fairchild employees. By 1968, Noyce and Moore decided it was time to go and set up their own company for the thrill of it.

They raised the capital, based entirely on Noyce's reputation, with one telephone call to Arthur Rock. Noyce invited his college, Grinnell in Iowa, to invest and Grinnell put up $300,000. Rock put up another $300,000 himself, and Moore and Noyce $250,000 each. The other six rebels also put up some money. Rock made 15 telephone calls and got 15 additional investors.

They called the new company Intel.

In 1970, Ted Hoff, a Stanford researcher working for Intel, figured out how to get all the circuitry of a computer on a single chip: the microprocessor. The marriage of transistor and the computer was consummated and the new age begun. By the mid-1970s, Intel had earnings of $100 million.

Figure 19 Gordon Moore (left) and Robert Noyce (right), at the founding of Intel, around 1960.

Intel soon totally dominated the semiconductor business. Moore and Noyce became richer than they had ever dreamed. The other six?

'If they kept their stock, they made out swimmingly,' said Moore with a huge smile.[11]

They lived Bill Shockley's fantasy. They directed the flow of the technology and made billions.

—<

Up the street, things were not going nearly so well. Shockley's outburst to Beckman not only triggered the great rebellion, but also soured his relationship with his backer. Shockley was supposed to get a new contract in September 1957, but Beckman was not returning telephone calls and was not visiting as often (or was not telling Shockley when he was in town). By December, Shockley still didn't have a signed contract.[21] Beckman missed appointments. Once, Shockley showed up ten minutes late in Los Angeles and found that Beckman had already taken the stage at a Chamber of Commerce presentation and would not be disturbed. Shockley sat in a side room for hours until Beckman decided to free himself.[15] Shockley called it 'very odd behavior.'

By March of 1958, Beckman had decided to renegotiate the terms of their arrangement. Shockley Semiconductor was still not making money. Both Shockley and Beckman were watching the rebels at Fairchild intently. Shockley's archives are full of newspaper stories about the rise of the company. By 1959, Beckman was so disturbed that he ordered a private investigation to see if Noyce and Moore were making use of Shockley Semiconductor corporate secrets. In fact, much of the preliminary work had been done at Shockley Semiconductor, but none of that was actionable. When someone leaves a company, they take wisdom with them and that is not theft. Beckman dropped the matter.[22]

What also is surprising is what Shockley did *not* learn from his experience. He was full of introspection, sometimes writing notes to himself. He acknowledged that all the rebels had deep respect for his scientific intelligence, but he concluded that he had spent too little time on human relations. People found it hard to read his thoughts, he said. He was too controlling, he felt. His impatience and irritability were communicated to his employees. 'Problems are primarily interpersonal,' he wrote in his journal.[23]

Still, he came to the wrong conclusion: Obviously, he had hired the wrong people. What he needed was a bright collection of scientists who knew how to take orders. As soon as the eight departed the company, Shockley was off to Europe. There, thanks to an educational system in which professors ruled classrooms like divine-right monarchs, he found the kind of employee who would do as they were told. Again, he managed to put together a first-class team of researchers. Shockley's ability to find talent was undimmed.

It never occurred to him that if a flock of wild geese are sent off in the wrong direction, it doesn't much matter how close the formation is.

Despite the obvious lesson from Fairchild that Noyce's business plan worked and his did not, Shockley did not drop the Shockley Diode. If an idea came from below it was rejected automatically, and to acknowledge that Noyce had been right would totally upset Shockley's theory of management.

By April 1960, Beckman had had enough of Shockley and bailed out. 'The management of Beckman decided to take this step so that the full financial and manpower resources of the corporation could be applied to the projected growth of its existing manufacturing divisions,' the official press release said. In other words, further efforts with Shockley Semiconductor would be a waste. The purchaser was Cleveland's Clevite Corp. The unofficial story was different.[19]

Forbes magazine described it as being like the act of a cuckoo (Beckman), who deposits her eggs in other birds' nests (Clevite) and lets the other bird worry about hatching them. *Forbes*' idea was half right. In nature, the cuckoo's chicks murder the other bird's babies and take over the nest. In real life, Shockley Semiconductor's eggs never hatched. Beckman treasurer George J. Renne,[15] who Shockley incidentally blamed for many of his problems, said the sale was 'one of the best things that has happened to us for years. It was losing between $750,000 and $1 million annually.... Shockley is a brilliant scientist, but brother, he's no manager.'[25] Clevite said it was delighted.

Shockley Semiconductor then had 110 employees. At the same time, Fairchild laid off more employees than that, going from 1,300 down to 1,100 because of a temporary slump in the business. Noyce said that with new technologies you can never tell how many people you need. Fairchild had been profitable for almost two years, and in less than a year hired the same number back.[26]

Shockley's relationship with Clevite was warm, but he was slowly withdrawing from his business, spending more time teaching at Stanford and more time doing other things. Six years later, the company was sold again, this time to International Telephone & Telegraph Co., which announced it was moving the firm to Florida (demonstrating how much they understood the business). Most of those remaining employees refused to go and the only sign left of William Shockley was his name on the library door. The company quickly disappeared, never having made a profit for anyone.

—<

Perspective needs to be added here.

In the late 1970s, the American Electronics Association published a genealogy of Silicon Valley. The table showed that virtually every company in the valley could show a line leading directly to someone who worked at and eventually left Fairchild Semiconductor. Fairchild became what writer Tim Jackson called the 'sycamore tree' of the valley. What was wrong with the chart was that it did not go back far enough. Everyone from Fairchild originally came from Shockley Semiconductor. Shockley's company was the seed of Silicon Valley.

Along with Fred Terman, Bill Shockley was the father of Silicon Valley.

Shockley lived long enough to see his child grow into an economic miracle without him. It was quite a sight: More people became richer in Silicon Valley, and became richer faster, than at any time in the history of the western world (with the possible exception of Amsterdam in the 17th century during the time of the Dutch East India Company, or Venice and its merchant fleet in the 15th century). In 1997, *Forbes* magazine listed Gordon Moore's wealth at $8.8 billion and named him as the fifth wealthiest man in America.[31] Of the top ten, five got rich from the industry Bill Shockley began or its software progeny. In the year before, Santa Clara county alone, constituting most but not all of Silicon Valley, earned more export dollars than any other metropolitan area in the United States: $29.3 billion, or fully 5% of the nation's total exports, exceeding both New York and Detroit, the next closest.[32]

Shockley earned none of that wealth.*

The valley also provided the intellectual and technological crucible for the greatest transformation in society since the Industrial Revolution in the late 18th century.

Shockley earned none of the credit.

—≺

Alas for Shockley, *Nemesis* was not done.

On a sunny Sunday afternoon in July 1961, Shockley, Emmy and Dick, who with his brother was spending the summer in Los Altos, decided to go out for dinner on the coast. Shockley drove, Emmy and Richard sharing the front seat of Emmy's Ford, with Emmy in the middle. As usual in the summer, the coast was wrapped in chilled fog, even if the rest of the area was sunlit. The road, Cabrillo Highway, or Route 1, hugs the coast south of San Francisco, and is treacherous even in the best of circumstances.

A station wagon driven by a drunk swerved out of its lane and hit the Shockley car head-on near the little fishing town of Moss Beach.[27]

Richard, then 13, was thrown to the pavement. He remembered the gritty taste of glass in his mouth. His father and stepmother were unconscious. Blood was everywhere. Both of them were flung into the windshield while the Ford's engine bolted back into the passenger cabin. Shockley's head injury was particularly serious, and he had a broken pelvis.[16,17] He came very close to dying.[28] Richard, the least injured, was treated and released.

Shockley was in hospital for a month; Emmy for six. The two of them were on crutches: he for a year, she for three. Emmy left the hospital in November. Shockley rented a Cadillac to whisk her to the Mark Hopkins hotel in San Francisco for an evening in the city. Then he took her to the Highland Inn in Carmel and they walked, she still on crutches, around Point Lobos. After two days on the Monterey Peninsula, they ended the idyll with dinner with May at Ming's.[29]

---

* Had Noyce lived long enough, he would have made it six out of ten. The others were Bill Gates, Paul Allen and Steve Ballmer, all of Microsoft in Washington, and Larry Ellison of the software giant, Oracle, of Redwood Shores, in Silicon Valley. Others in the top 25 included Michael Dell of Austin (Dell Computers) and William Hewlett. David Packard, who died in 1996, also would have made the list.

In December, he was still trying to touch his toes with his legs straight and bending his leg to his chest in the pool. By 20 December, he could do 26 push-ups.[30]

They never recovered completely. Shockley never climbed again. Emmy could not close her right hand.

How much else they were hurt, no one can say. After Shockley's death, Fred Seitz suggested that the head injuries were responsible for what came after. 'I am inclined to believe that the residual effects of his near-fatal accident cannot be ignored in evaluating his activities in later years,' Seitz wrote to *Science*.[28]

What is clear is that Shockley wasn't the same again. Moore said the accident aged him 20 years.[10] Shockley told his son that he and Emmy were less cheerful after the accident, partly because of their constant pain.[35]

His daughter doesn't think the accident affected what happened next. His business had collapsed and he had nothing to keep him interested, nothing to throw himself into. He was past his creative peak. He had failed and he hurt.

A distraction was not long in coming, and it began in as lovely and gentle a place as anyone could find.

CHAPTER 10
'Three generations of imbeciles are enough'

The ambience would have eluded him.

One day in early May 1963, Bill Shockley was in an archetypal mid-Western college town, about as close to the heart of America as he ever would likely be. The weather was lovely, the prairie was ablaze with spring and he was on a college campus on a warm, sunny day. The crab apples already were blossoming, and the elms, red cedar, spruce, and hagberry trees wore their bonnets of deep green leaves.

Shockley and his escorts, a student and a professor from Gustavus Adolphus College, made the trip to St Peter, Minnesota, southwest of the Twin Cities airport, in a little more than an hour. In the middle of town, the car turned right, past the large park with its bandstand and playground, up the street past lovely old Victorian houses, brick homes with ivy-covered walls, and white frame houses behind fences, all blooming with vibrancy and color. At the crest of the hill, the road circled Gustavus Adolphus College. There, Shockley joined the other Nobelists and the King and Queen of Sweden for the dedication of the new building.

Gustavus is one of the small denominational colleges (student body: about 2,000) founded in the middle of the 19th century to give working and middle class students a quality liberal arts education in a Christian atmosphere. It is Lutheran and Swedish. Gustavus people are very quick to proclaim this affiliation, lest you – God forbid – confuse it with St Olaf in Northfield, about 40 miles east, which is Lutheran and *Norwegian*. In the early 1960s, the college decided to build a new science building and, to capitalize on their Swedish connections, contacted the Nobel Foundation for permission to name the building after Alfred Nobel. The foundation agreed. The center hall of the new building would become a small museum for the Nobel Prize. At the suggestion of Peace Prize winner Ralph Bunche, the

foundation also agreed to let Gustavus hold an annual Nobel conference, the only one sanctioned outside Sweden and Norway. That was
scheduled for two years hence. The college thought the conference
and Nobel Hall would be wonderful for its somewhat insular student
body, and would provide great publicity and a fine recruiting tool for
faculty and students alike.

Twenty-six Nobelists, including Shockley, came for the dedication.

Because the college was unsure that the media in Minneapolis and St
Paul would bother to make the hour drive to St Peter – even on a lovely
day – it scheduled a series of press conferences at the airport as the
Nobelists arrived in clusters.

One of the regulars at the succession of conferences was a young science writer from the Minneapolis Tribune, Victor Cohn. Cohn, one of
the few science specialists there, soon found that he was one of the few
with intelligent questions for the Nobelists. The other reporters,
mostly from radio and television stations, had no idea what you ask a
Nobel Prize Laureate in physics or chemistry or even medicine, so
Cohn dominated the first press conference, asking most of the questions. Like many specialist journalists, he had little sympathy for his
technically challenged colleagues who were slowing things down when
he was on deadline, so after the first conference, he decided to commit
a little sabotage. He stationed himself to the side of the room, so that
to answer his questions the Nobelists would have face him, in profile to
the cameras – which television stations hate – and from the micro-
phones – which muffled the sound. It was more subtle than pulling
their power cords.[3]

This was still the height of the Cold War and mutual atomic terror, so
Cohn asked Shockley (in profile) what he thought the chances were of a
nuclear war. Shockley responded. 'Fifty-fifty, I think.' Then he paused a
second. 'But if there is nuclear war man would at least have to begin to
control his own genetics. I think the present situation in the civilized
world is anti-evolutionary. The people who reproduce in the largest
numbers may be far from the most competent. The more competent
people practice birth control and have smaller families. If there were a
nuclear war, there would be so much genetic damage that man would
then be forced to plan populations – yes, control breeding, that's what it
would amount to. If we began sensible population measures now, it
would make nuclear war less likely.'[34]

Cohn thought that an interesting response and used the quote in his story. Shockley, who liked a Manhattan or two, admitted later that he had had a few drinks on the plane, which may have had something to do with the somewhat indiscreet answer.* He claimed he had never thought of the question that way before, but was himself struck by what he said. He was afraid also that Cohn or his editors would blow up the story and the headline would say something like NOBEL LAUREATE THINKS NUCLEAR WAR WOULD BE A GOOD THING. Cohn, a superb reporter, played it straight, as did his editors.[4]**

The next February, he followed his thoughts more fully at an invited speech before the Planned Parenthood League of Alameda County (the Berkeley–Oakland vicinity). Shockley warned that if 'exponential growth of world population' continued unchecked, it would lead to starvation. On the other hand, he said, if population growth is controlled, there is a danger of making life so good that humanity could suffer evolution in reverse. Darwin had described a scenario in which the fittest survive; Shockley was concerned about a world in which not only do the unfit survive, they are fruitful and multiply stupidly.

'Those very things which are now giving us our highest standard of living may have an anti-evolutionary effect,' he said. 'In all past periods of civilization, danger of starvation and death from other violent causes existed. Today our high standard of living may result simply in a predominance of the people who can produce the most offspring. If this criterion alone is selected for determining the future characteristics of the species, it is extremely likely that this would have a very adverse effect.... Those of us who care about this future should urge that the problem be given one of the highest priorities for scientific study by our ablest scholars,' he concluded.[35]

Shockley was also invited to be a speaker at the first of the Nobel Conferences after the dedication at Gustavus two years later, in January 1965.

This time the town was in the depth of the Minnesota winter, with temperatures below zero and all the lovely Victorian houses – the ivy

---

* There is no reason to believe drinking was an issue.
** Shockley said the idea popped into his head when Cohn asked the question. Embarrassed at being the instigator of what followed, Cohn, who ended an illustrious career at the *Washington Post*, asked him not to tell anyone.

and brick mansions, the frame homes – and all the trees huddled stoically against the gelid prairie blasts. The theme of the conference was 'Genetics and the Future of Man.' Shockley was one of three Nobelists asked to speak: the other two were Polykarp Kusch, a winner in 1955 for his work on measuring atoms, and Edward Tatum, a biochemist who won the 1958 Nobel for his work using bread mold to show how genes regulate chemical events. Another speaker was the ethicist Paul Ramsay. Why Kusch and Shockley, physicists both, were invited to a genetic conference remains a mystery, although Kusch was on the committee that organized the conference and may have thought he had something to say. Eight thousand people showed up for the conference, held mostly on the second floor of Alumni Hall, in a large rectangular ballroom.

Shockley would call the meeting the 'turning point of my life.'[36] Indeed it was.

He described himself as a 'non-specialist'[37] concerned about the quality of human life. His concern, he said, came from personal experiences and admitted 'these personal experiences do not qualify me as an expert in the fields of genetics and sociology, and my credentials are not of comparable standards with other speakers of this symposium.'

His concern for the future began during his wartime tour in India, he said. With eloquence that would be sorely missing from later speeches – rhetorically it was probably the best speech of his life – he told the Gustavus audience of the crowding and the sheer mass of humanity he saw. He described the villages on the Bengali plain – how clean they were because the Indian villagers scavenged anything that would normally be litter or garbage in a western city for its utility. They even retrieved animal droppings to be dried and used as fuel.

When he returned to the States he read a booklet on the population explosion and the dangers of starvation, he said. He learned that it takes seven calories of grain to feed an animal to produce one calorie of food for a human. That works well in the US, where half the calories come from plants and half from animals, and where there is plenty of both to go around. In India and China, almost all the calories are from vegetables and there is no margin of error – if a crop fails, people starve.

'On the basis of these ideas, I at first felt that I would not be in favor of sending food to relieve a famine in India. To do so would simply make the situation worse between that famine and the next.' When a famine struck a few years later, however, Shockley found that the morality of sitting on a grain surplus in the US while people were starving in Asia outweighed his cool Malthusian reasoning. Things, he admitted, were complicated.

'I feel it is of importance to think about the problems and provoke discussions so that wiser decisions can be made when it inevitably becomes necessary to make them.' He explained how the Earth's population was growing exponentially. He pointed out that some more advanced countries, such as Denmark, Sweden and Japan, managed to slow their population growth with contraception and abortion; the latter, when done properly and legally, being a lot safer to a woman than childbirth.

So far, Shockley had said nothing shocking. He hardly was the only person lamenting over-population – most in the audience probably agreed – and the issue would grow in importance in later years. Then he went on.

In what would become one of his most often-told stories, he described an incident in which a delicatessen owner in San Francisco was blinded by acid in an assault. The story grew in detail with later telling, but in this speech, he told the Gustavus audience that the young assailant had been hired by an emotionally unstable woman with a grudge against the delicatessen owner. The young man was one of a family of approximately a dozen illegitimate children on welfare. 'This brought home to me the possibility that if we had a situation in which an irresponsible individual could produce offspring at a rate which might be four times greater than those of more responsible members of society, this was a form of evolution in reverse.' Shockley did not mention in this telling that the young man was black, the shop owner white.

He listed three things he felt threatened humanity: nuclear war, famine, and finally 'genetic deterioration of the human race through lack of elimination of the least fit as the basis of continuing evolution.'

Those multiplying the most were the less intelligent and intelligence was largely inherited, Shockley said. He described one experiment with mice to support that contention. A group of laboratory mice was selected for either their speed of learning new tasks or their slowness, and each group was bred separately. By the ninth generation, the differences in mouse intelligence between the groups were distinct and obvious. One group consisted of rapid learners; the other group consisted of animals that were dumb even for mice.

He said that the genetic component of intelligence was not one gene – there was no single gene for genius or stupidity – but was the action of many. He was well aware of the obvious anomalies that seemed to argue against a genetic component to genius. He mentioned Leonardo Da Vinci, who was the illegitimate son of a nobleman and a peasant girl, the only outstanding offspring of either family. How could such non-distinguished families produce one of history's great geniuses? To Shockley, no one familiar with statistics would be surprised at Leonardo's origin. Because of the multiplicity of genes ('about ten thousand billion possible offspring can result from making the random selection from the twenty-three pairs in the mother and the twenty-three pairs in the father'), the exceptions just prove the rule. You would expect a Leonardo a predictable number of times if you had babies often enough. Sooner or later, the genes would fall that way. The laws of probability are not suspended by sex.

'There is no reason to doubt that the genetic aspects of intelligence are governed by such probability laws,' he concluded. 'There is no reason to doubt that genetic probability laws apply to human intellectual and emotional traits.'

Shockley at this stage was certainly no geneticist, and he didn't claim to be one. He was, however, one of the greatest living experts on the use of statistics for analyzing human behavior and conditions. If, in the ensuing debate, his critics would show a malicious lack of respect for Shockley's knowledge of genetics, he felt the same way about their ability with statistics. Once he reduced the problem of intelligence and genetics to the laws of probability (which clearly would not be an acceptable reduction to many), he felt he could see things they could not.

What to do about the unfit? Sterilization and abortion. Shockley was not alone in that sentiment. Tens of thousands had already been sterilized by the US government in the first half of the century. In a celebrated court case (*Buck v Bell*, 1924), the State of Virginia ordered the sterilization of a mentally retarded woman to prevent her from having children. The great supreme court justice Oliver Wendell Holmes wrote: 'It is better for all the world if instead of waiting to execute degenerate offspring for crime, or let them starve for their imbecility, society could prevent those who are manifestly unfit from continuing their kind.... Three generations of imbeciles are enough.' (Shockley added that seriously retarded people were not the ones he was talking about; they rarely reproduce, and retardation usually is not genetic, he said, not entirely accurately.)

He ended his speech, totally unaware of what he had just done to himself.

Compared to what came later, that was fairly mild. Race never came into the discussion. Still, it upset several people in the audience, a few of whom protested to Gustavus president Ed Carlson. Carlson responded that a Christian liberal arts college was supposed to raise profound social issues and he offered no apology for Shockley.[38]

As Shockley would soon learn, Carlson's courage would be rare.

After several days at Gustavus, chatting with students and faculty and generally being charming (and he could be quite charming when the mood struck), he returned to Stanford and a new life.

With his company, now owned by Clevite, barely hanging on, Shockley had not a great deal to do. Linvill arranged a 'chair' for him, a small endowment given to Stanford by a founder of Ampex, an electronics company famous for sound recording, to honor Alexander Poniatoff, another founder. The money wasn't much, about $1,000 at first, but Linvill convinced J. M. Pettit, the dean of the school of engineering, that it was a good way to get Shockley in the door. He could work part-time while doing what he needed to do for Clevite. The physics department was not unhappy to lose him. It was concentrating on a different field of work using Stanford's first linear accelerator to study the structure of the atomic nucleus – and had won a Nobel Prize for the work.* Shockley's interests didn't mesh.[39]

Pettit wrote Shockley that the university hoped he would some day be able to take a full-time position and would wait until he worked out his responsibilities to Clevite. Meanwhile, Pettit wanted him to run a graduate seminar in solid state electronics, and to be around when other professors or students gave presentations in that field. That would be about once a week, Pettit said. Also, the school wanted him to take over as an informal advisor to some graduate students. Additionally, Shockley's appointment was 'at large,' meaning he could float around the school. Solid state research was spread across two departments, electrical engineering and materials science, 'and we know that you would strengthen the work in both these areas. I am sure you would have useful

---

* Robert Hofstadter, 1961.

connections with other parts of the university as well, such as physics, chemistry and applied physics,' Pettit wrote.[40]

Shockley couldn't turn down such an offer. He believed – rightly – that he was a superb teacher, and Stanford was providing a professional home, an identity.

—<

Shockley was no longer able to climb cliffs because of the accident injuries, but he had found another hobby, one he and Emmy could share and love – sailing.

During the visit to Woods Hole in 1957, while his company was self-destructing, Shockley rented a small sailboat called a Hobie Cat, with a small keel, one sail and a steering lever. Shockley got a compass, a map, a protractor and a packed lunch, put Emmy and young Dick in the boat and sailed off into the bay. He seemed, Emmy remembered, to know exactly what he was doing.

'You must have done a lot of sailing,' she commented. 'You seem to know so much about it. I didn't know you sailed.'

Shockley hesitated.

'You have, haven't you?' she asked.

'Well,' Shockley admitted, 'I've read some books.' He had never piloted a sailboat in his life. That didn't stop him from taking his wife and a son out into the bay on one.

Emmy loved it. Next day, she and Dick went out alone. She proposed they practice so they could take Shockley for a ride. Neither of them had been in a sailboat before either. They practiced all afternoon and then sailed up to a surprised and immensely pleased Shockley who was standing on the dock and whisked him off over the water with them.

When they returned to California, they decided they needed their own boat and purchased a 25-foot bay sailor called the *Fandango*. They cruised up and down the wind-ridden San Francisco Bay out of Redwood City, sailing in races and winning several trophies. Once, they even broke their mast in the wind.

After the accident in 1961, they found that the *Fandango* was simply too much; Shockley was hobbled and Emmy still was on crutches. So Shockley bought a larger boat, one that could be rigged from the cockpit. Emmy named the 34-foot ketch (classed as a Chesapeake Bay sharpie) the *Sly Sharpie*, because its centerboard could be raised so that the boat could slide over the mud banks in the shallow southern end of the bay while others ran aground.

Shockley was involved in at least two sailing accidents, injuring himself slightly once and once getting sued, but he and Emmy adored sailing. So it seems did his students. The Shockleys invited them to lunch sails. Emmy made a concoction of clams covered with a mixture of tomato juice and clam juice. That served as an appetizer. The main course usually consisted of Kentucky Fried Chicken and biscuits. She served a tart or pie for desert. 'They loved to go out,' she said.[41] They probably didn't come for the food.

He quickly acquired a freshman seminar, taught twice a week, which he limited to 16 students. The seminars often met at the Shockley home, in the family room, where Shockley set up a blackboard. Emmy served snacks. Shockley taught the students problem-solving, attempting to teach them how to conceptualize problems as he did, not to take the obvious and usually more time-consuming route, but to look at problems through different eyes, something he could do better than almost any one. Twenty years later, the technique would be known as 'conceptual blockbusting' and be taught widely at colleges. One academic quarter he had 32 qualified students apply for the seminar. He chose 16 at random. He followed the 16 he admitted to the seminar and the 16 he had to exclude through their Stanford careers. His class did markedly better than the ones who did not get in, and Shockley was sure – and many of his students agreed – that it was because of the methods he taught them when they were freshmen.[41]

In 1963, Dick, then 16, whose relationship with Jean had deteriorated into something resembling hatred, announced he wanted to move to California to be with his father. May was thrilled to have a grandson around to spoil. He finished high school living with them.

In 1966, Shockley took advantage of one of the great perks of a Stanford faculty appointment: he, Emmy and Dick moved onto campus in the 'Faculty Ghetto.' Only permanent faculty and the highest levels of the administration can live there. They lease land from the university, but the houses belong to them. If they leave, or die, or if there is a divorce and the Stanford professor leaves home, the house must be sold to another qualified faculty member. Placed in the lush, hilly part of the campus, filled with palms, eucalyptus and flowering trees and gardens, the Ghetto is one of the more beautiful residential areas in the locality – or any place else for that matter. Each house is different: some ordinary, some striking. Shockley and Emmy chose a conventional brick-faced Cape Cod* on Esplanada Way, within walking distance of the community swim club. The house, shaped like a backward 'L', had a living room, family room, large kitchen and an attached garage. Shockley moved his office into one of the three bedrooms and some of his files into the garage.

The joy of the house was the backyard. Shockley built birdbaths and a deck. He and Emmy had most of their meals out there, weather permitting, and Shockley would spend evenings watching television or listening to tapes. He and Emmy sat hooked into headsets so that the neighbors wouldn't be disturbed. He fed the birds and the squirrels (getting to know several on a first-name basis) and continued his gardening.

They ate out often, usually at the Swiss Chalet in Palo Alto.

May, now nearing her 90th birthday, moved into a federally subsidized home for the aged in Palo Alto when her half-sister, Lou Vee, died. She was healthy and did not need much assistance. The Shockleys had dinner with her frequently and, mind still keen, she followed the exploits of her dear famous son avidly. He seemed now to find her tiresome, but they kept regular contact by phone or visits. She was only a mile away.

He set up a program at Wilbur Jr. High in Palo Alto on problem solving and, along with the Stanford School of Education, applied for a federal grant to develop the technique.

Shockley joined the Bohemian Club, an all-male group of prominent businessmen and political leaders, who owned a large, exceptionally beautiful, campsite in the redwoods of Marin County, north of San Francisco. Shockley happily spent a week every summer for the rest of his life at the Bohemian Club encampment, even participating in the skits that often required members to dress in drag. There was no need to worry: the guy in drag next to you could easily have been the Secretary of State.

Alison was living quietly in Washington with her husband. Shockley visited them when he was in town, but was not close beyond that. One or the other might call on a birthday. Dick lived in relative peace with his father and Emmy. Bill had gone off to college, but was unhappy there and left for New York City. 'I wasn't doing anything that he could admire at all. I don't even remember what I was working at. I don't have a

---

* A single story house with a steep roof.

degree. I had gone there and became a hippie-beatnik type. I learned to play the guitar. I had experiences on the street with people that Jack Kerouac wrote about.'

One day in 1964, Shockley came to New York and dropped in on Bill, suggesting the two should spend some time together. Bill had a copy of *Mechanical Engineering* magazine around, and assuming it would be something to engage his father, he brought it to Shockley's hotel room. The magazine had a section on mechanisms, 'weird ways of things being connected to transfer this motion into that motion.' They sat around the hotel room the first day discussing the devices.

They did it again the second day. By this time, Bill's head was swimming. He asked Shockley if they could take a break from physics and just go see a movie or have a beer or something. 'Teaching is what I do, and if you want to do it, I'll stick around,' Shockley told his son. 'If not, I'm going to go back to California.' Shockley left the next day.

Bill never saw his father again.[42]

—<

Early in July 1965, an editor at *U.S. News & World Report* magazine contacted Shockley to set up an interview. He had read about Shockley's Gustavus speech and thought it would be worth expanding. A team of *U.S. News* writers interviewed Shockley on 24 July.

He expanded on his Gustavus speech. He quoted US labor secretary Willard Wirtz as saying that a disproportionate number of the unemployed *seem* to come from large families. Wirtz lamented there was no solid research to prove if this was really so and he told Shockley he hoped what he said would trigger such research.

'In other words, we're not finding out if this is true,' Shockley said. He added that every time he asked geneticists if 'large improvident families with social problems simply have constitutional deficiencies,' the geneticists invariably reply that they don't know and were not about to find out.

Shockley said he was convinced that some people are born with physical deficiencies in their brains, usually in the frontal lobe, just as if they had lobotomies, he said. He was worried that this was happening to too many Americans. He then returned to the case of the delicatessen owner, adding facts, including the race of the assailant and victim. The youth who threw the acid was one of 17 children of a mother who had an

IQ of 55 and could remember the names of only nine of her offspring. The probable father died in prison after a murder conviction. She should not have reproduced. Shockley said he didn't know if she was 'an isolated statistic,' but he feared she was not.

'There are some who deny these dangers on genetic and statistical grounds. But I have little confidence in the objectivity of their reasoning or the reliability of their optimism,' he said.

He now turned to intelligence and heredity. He was convinced of a strong genetic component in intelligence, quoting some conclusions of a twin study that showed that identical twins differ in IQ far less than do ordinary brothers and sisters. The twin studies had been done of identical twins separated at birth and raised in different homes. The sample was small, he said, about 100 children, but he found the results intriguing. He asked that more research be done using control groups of abandoned children.

Then one of the *U.S. News* interviewers asked: 'To what extent may heredity be responsible for the high incidence of Negroes on crime and relief rolls?'

'This is a difficult question to answer,' Shockley replied. 'Crime seems to be mildly hereditary, but there is a strong environmental factor. Economic incompetence and lack of motivation are due to complex causes. We lack proper scientific investigations, possibly because nobody wants to raise the question for fear of being called a racist. I know of one man who is writing a book in this area, and I'm not sure he'll finish it because the subject is so touchy.'*

Shockley then stated that while the distribution of IQs among African-Americans (a term he would not have used in 1965) includes people of superior intelligence, African-Americans *as a group* have a mean IQ 15 points below the mean of whites. He pointed out that this was one standard deviation from the mean.

'How much of this is genetic in origin?' Shockley asked. 'How much is environmental? And which precise environmental factors are to blame? Again, a "controlled" program of adoptions might give answers.' He gave no details of how that would be done.

'Actually,' he said, 'what I worry about with whites and Negroes alike is this: Is there an imbalance in the reproduction of inferior and superior

---

* Probably Arthur Jenson at Berkeley.

strains? Does the reproduction tend to be most heavy among those we would least like to employ – the ones who would do least well in school? There are eminent Negroes whom we are proud of in every way, but are they the ones who come from and have large families? What is happening to the total numbers? This we do not know.'

Shockley said that part of the reluctance of the scientific community to answer these questions is the distaste that people have in thinking that human beings are subject to the same laws of nature as other animals, that somehow we are above all that. It's unnerving to them, he said. In a comment that eerily foretold later discussions, he pointed out the number of families on welfare who were producing second and third generations of welfare recipients. Instead of cutting them off from aid, as later politicians did, Shockley recommended sterilization.

He did not say in the interview whether he felt this should be voluntary or involuntary, although later he made it clear that he did not favor forced sterilization. He also supported abortion. Any program addressing the problem should not be based on race or economic class ('Poor people can be quite gifted'). Good breeding should be based on the genetic material of a family. 'We need more Lincolns, not fewer.'

'Several eminent intellectuals have discouraged me from publicly expressing the ideas we have talked about,' he added. 'They feel the uninformed and prejudiced might react badly. But I have faith in the long-term values of open discussion.'[43]

Several things stand out about the interview. First, he had obviously read and talked about the subject beforehand, including correspondence with a number of people he considered experts. He was spending considerable time on this topic by now. Second, he accurately described the feeling of most scientists about the subject: that little reliable research had been done and there were strong currents discouraging it. Third, he knew exactly what he was doing, even if he did not know the price he was about to pay.

Shockley's arrangement with the magazine was that he would have final approval of the manuscript. Now that arrangement is considered unethical in journalism, but was not then. When he saw a copy, he decided he needed to make sure of his footing.

Before the interview was published, on 22 November, he sent a letter to Robert Lamar, the science writer at the Stanford University News Service who would likely handle the press release and queries from reporters, explaining his position and his thoughts since reading a transcript of the interview. He told Lamar he felt competent to discuss genetics. He had learned from his experiences in the war, however, that sometimes a non-specialist can play 'a significant role as a team member.' That's what he thought of his role.[44] He was the Nobel Laureate. He would be the lightning rod. The magazine published the article.

The reaction was immediate and came from right up the street. Shortly after *U.S. News* published the interview, *Stanford M.D.*, a magazine published by the public relations office of the Stanford Medical Center, reprinted it. Stanford's illustrious faculty of genetics responded. The signatories included Joshua Lederberg, winner of the Nobel Prize for medicine in 1958 for his genetic work. They protested the republication of the interview, calling it 'pseudo-scientific justification for class and race prejudice.' They called Shockley's comments 'so hackneyed that we would not ordinarily have cared to react to it. However, Professor Shockley's standing as a Nobel Laureate and as a colleague at Stanford, and now the appearance of his article with a label of Stanford medicine, creates a situation where our silence could leave the false impression that we share or even acquiesce in his outlook, which we certainly do not.'

The faculty said that Shockley's suggestion that serious research be done on genetic factors in social maladjustment and 'certainly the need for more creative imagination than we now observe in planning social welfare and in education' were outweighed by the 'mischief' of distorting social responsibilities. 'Too many people will seize any excuse for these purposes. The plain fact is that we do not know the answers to his provocative questions, and in our present day context it falls between mischief and malice to make such a prejudgment in his terms.'[45] In other words, as Shockley said in his interview: 'we don't know the answer to that question and we don't want to find out.' They called his solutions 'totalitarian.' How it was pseudo-science, they didn't exactly say. They added that Shockley overlooked the fact that society changes much faster than heredity, and that Shockley had ignored the need to improve medical care, education and the economy to create incentives and 'useful careers for the whole wonderful variety of humans.'

Shockley was shown the letter before it was published so that he could reply, and so that the magazine could publish both letters simultaneously. Shockley talked to several people about how to respond, including, apparently, Harvey Brooks, chairman of the National Academy of Sciences Committee on Science and Public Policy. According to

Shockley, Brooks said: 'Ignore the question of race; otherwise you are simply guaranteeing yourself against an objective audience.'[46] Shockley took his advice – this time.

In his response, Shockley asked if using such loaded labels as 'pseudo-science,' 'hackneyed' and 'mischief' added to free inquiry or whether their 'totalitarian and dogmatic posture' was an attempt to dictate permissible channels of thinking. The whole context of his interview, he maintained, was 'let's ask the questions, do the necessary research, get the facts, discuss them widely – then either worries will evaporate, or plans for action will develop.' He said he admired intellect, even when he disagreed with it, and 'deplored feeblemindedness (especially an IQ of 55 with seventeen children).'

'Here I disagree with the genetics faculty; I cannot in good conscience apply the word wonderful to the feebleminded "variety" of humanity.'

His response to the charge that it was pseudo-science was a long paragraph with scientific citations purporting to support his position, and at least one quotation from another scientist also suggesting sterilization as a valid option. He said he had corresponded with other scientists, including geneticists, who agreed with him but were afraid of publicity. They denounced to him what they felt was a lack of integrity and objectivity of the general scientific community, he said. His response to the argument that the results might be misused to evil ends was that the genetics faculty was underestimating the integrity and intelligence of the American people.[47]

The Stanford geneticists were also the first to underestimate Shockley's tenacity and his ability to do his homework. They apparently assumed that since he was not a geneticist he could have no idea what he was talking about and he would be an easy target. Others would make the same mistake.

Shockley then decided to take his argument to the big time: the august National Academy of Sciences. As a member, Shockley had the right to speak and make motions, and that's what he did at the NAS meeting at Duke University on 17 October 1966.

Shockley asked the NAS and other national scientific organizations to rise above the argument of racism and study what effects genetics had on the problems in America's slums. Shockley said too many people feared that the research would inevitably lead to 'intelligence distributions of ethnic minorities in general and American Negroes in particular.' Consequently, very little work was being done. Scientists were busy making declarations of extreme views on both sides with little or no data to support the views. Yet, he said, there are tantalizing hints.

He described one study in which 47 Oregon babies of schizophrenic mothers were placed in normal environments after birth. Half showed some 'mental disadvantages,' he said, including 'mental deficiency, schizophrenia, criminality and discharge from the armed services for psychiatric and behavioral reasons.'

While race was not a large factor in his assertions until the *U.S. News* interview, it was now. He proposed a mathematical 'H-index' to act as a yardstick for figuring out the genetic ancestry of individuals, how much white blood there was in African-Americans, for instance. The index – which he dropped after this proposal – would provide a base, an 'objective benchmark' for research. Even other scientists who essentially agreed with his proposition that intelligence was largely genetic would stop dead in their tracks rather than supporting something like that, but they would be showing a sensitivity Shockley now lacked.

He urged that the academy set up a summer study group to seek new ways to utilize 'scientific imagination to reduce the environment-heredity uncertainty as related to the problems of the city slums.' It would work, he said, only if the scientific establishment had the courage to ask the questions and not be afraid of the answers.

'I find I cannot in good conscience walk away from this challenge,' he told the academy, 'and I feel that I have a greater obligation to face it than would a life scientist whose professional risk would be greater than mine.'[48]

The president of the NAS, Shockley's old companion, Fred Seitz, asked a committee of geneticists to take up Shockley's question. A year later, the committee released its report.

'With complex traits like intelligence the generalities are understood, but the specifics are not,' the committee reported. 'There is general agreement that both hereditary and environmental factors are influential; but there are strong disagreements as to their relative magnitudes – which is another way of saying that the evidence is not conclusive.... It is unrealistic to expect much progress unless new methods appear.... To shy away from seeking the truth is one thing; to restrain from collecting still more data that would be of uncertain meaning but would invite misuse is another,' the committee said. 'It is contrary to evidence that social problems such as poverty, slums, school dropouts, and crime are entirely genetic. There is surely a substantial and perhaps overriding environment and social component.'

Then there was the issue the committee noted of the results of such experiments. If it were proven beyond a doubt that social maladjustment was hereditary, what would we do about it? This was the 1960s, well before *Roe vs. Wade* and wide use of the birth control pill. Artificial insemination was still new and had legal problems. Even if the decision were made to take the eugenic approach, legal restrictions on techniques such as abortion would prevent implementation.

'We question the social urgency of a greatly enhanced program to measure the heritability of complex intellectual and emotional factors,' the committee concluded.[49]

Seitz said later the motion to launch the research was defeated because there was fear it would turn into a 'white vs. black issue.'

At the time he said, 'There is a strong feeling that in the current circumstances in which the social issue is so predominant – we're trying to find our way in equal opportunity – that it would be almost impossible to carry out reasonable research... that it would not be misunderstood.'[50]

—<

Shockley was now espousing a theory called eugenics, the bastard offspring of Charles Darwin's theories of evolution. It wasn't Darwin's fault.

Eugenics has a long and distasteful history. It's not possible to have any rational discussion of it without running immediately into that mire. The theory and the movement it engendered began with one of Darwin's cousins, a British statistician named Francis Galton, who coined the word in 1883. 'Eugenics,' Galton wrote, 'is the study of the agencies under social control which seek to improve or impair the racial qualities of future generations either physically or mentally.'

One popular extension of Darwinian selection – to which Darwin himself did not subscribe – was the feeling that what happened in the natural world was mirrored in human society and economic classes. The upper classes, the leaders, were the most fit to survive and did; the lower classes clearly were in that position because they were inferior, and many of them did not survive. To some extent, this rickety theory provided *post hoc* validation for the British class structure and American slavery: those bright and strong enough to catch, sell and keep slaves would, while those unfit enough to be caught, sold and kept would be slaves. This social Darwinism intrigued Galton for a time. Then he thought there was something more important afoot. Civilization, which meant social programs, health care, economic advances, and perhaps even charities, were providing a safety net for the less fit. No longer were they being eliminated or even discouraged from reproducing. Civilization had interfered with nature's laws and the unfit were surviving. Government and society must intervene to tilt the balance back.

Galton's theories became particularly popular in the US after the turn of the century because the country was undergoing radical change, particularly with the tide of immigrants and the social and economic changes that ensued. The immigrants passed under the watchful eye of a statue of a beautiful woman holding a torch and standing on a pedestal. Engraved on the pedestal was a poem urging the tired, the poor, the huddled masses of the world to enter America's golden door. Millions of the tired and destitute were accepting the offer, huddling in steerage on almost every steamer crossing the Atlantic. The opinion that this was not a good thing for America was widely held, particularly by those of Anglo-Saxon lineage who had been there long enough to forget that their ancestors did the same thing. These anti-immigration forces found allies in nativist and racist organizations, including the Ku Klux Klan.

One supporter of eugenics was a former Harvard instructor and assistant professor of zoology at the University of Chicago, Charles Davenport. In 1904, Davenport founded the Station for Experimental Evolution (SEE) at Cold Spring Harbor on Long Island Sound. Davenport's work centered on breeding animals and plants. In 1910, Mary Harriman, widow of the railroad magnate Edward Henry Harriman and mother of W. Averell, the great liberal politician, agreed to fund an expansion of Davenport's work into human evolution. The Carnegie Foundation also contributed to funds.

Davenport wrote a text, *Heredity in Relation to Eugenics*, which was used in colleges around the country until the 1930s. He concluded that any time a family showed a high incidence of a given characteristic, it was inherited. He wrote that often only one gene was responsible for such things as Huntington's chorea, albinism and hemophilia. Other traits, such as mental prowess and behavior, required more than one gene. He wrote there were genetic components to alcoholism, insanity, epilepsy, criminality, 'feeblemindedness' and pauperism. Sometimes, he was right, as modern genetics science has borne out.

Davenport felt that race determined behavior. He was a raving anti-Semite – 'hordes of Jews' were coming, he warned. His book called for securing the 'best blood' for America, not the detritus of central Europe. 'Man is an organism – an animal,' he wrote, 'and the laws of improvement of corn and of race horses hold true for him also. Unless people accept this simple truth and let it influence marriage selection, human progress will cease.' He warned that unless immigration stopped Americans would have darker skin; grow smaller and more emotional; and become more 'given to crimes of larceny, kidnapping, assault, murder, rape and sex-immorality.'

Again, this was no small movement of society's malcontents mumbling to themselves on the street; eugenics was taught in most American colleges until the war. The movement influenced anti-immigration legislation that finally led to the slamming of Miss Liberty's golden door. Respectable academics, including well-known scientists at famous institutions, supported it.

Davenport used Mrs Harriman's money to found the Eugenics Record Office (ERO) and brought in Harry Laughlin, a Princeton-educated instructor at Northeast Missouri State Teachers College, to run the office. ERO became the clearing house for data on the inheritability of epilepsy, and such things as eye color, hair color and skin pigmentation. The main function of ERO was to gather data on families on forms called 'Record of Family Traits.' The office also amassed data on entire high school graduating classes and college students, a huge collection of material.

Davenport began to concentrate less on how 'good stock' could increase and more on what he called 'negative eugenics.' Fifty years later, Bill Shockley would call this 'dysgenics.'

Laughlin believed the only solution to the problem of down-breeding was government intervention. He drafted model state laws calling for involuntary sterilization of criminal elements, alcoholics and the retarded. Between 1907 and 1928, 21 states passed laws based on Laughlin's model, including the one in Virginia that Oliver Wendel Holmes defended. Twenty thousand Americans were sterilized against their will.

The success of the movement attracted the attention of the National Socialists in Germany, largely because it fit well with the racial theories of their leader. In December 1936, Laughlin was given an honorary doctorate from the Nazi-dominated University of Heidelberg for his eugenics work. He was thrilled. He drove to the German consulate in Manhattan from Cold Spring Harbor for the award ceremony, which he called a 'personal honor.'

By 1939, the Nazis translated eugenics into their 'euthanasia' policy, killing thousands of undesirables, including Jews, gypsies and homosexuals. The stench of what was happening was so great that the Carnegie Foundation pulled its support for the eugenics office. A group of geneticists organized itself in an assault on the movement, a rare event. Until then, the scientific community had said little, essentially letting Davenport and Laughlin go unchallenged. The ERO closed the same year.*

Eugenics in Germany evolved into the abomination of genocide. It would be nice to say that Laughlin and Davenport lamented what had happened to their 'science.' Alas, that was not true. Laughlin died in 1943 and Davenport in 1944. Both knew what was happening. Neither said a word.[51]

That was the intellectual company Bill Shockley elected to keep. His opponents did not intend to let him forget it. It became almost impossible for most critics to mention the word 'eugenics' in one sentence without using the word 'Nazis' within three sentences of it. Any honest scientific debate on the underlying theory was very difficult, or even impossible.

Yet Davenport and Laughlin were not the only company Shockley was keeping. One speaker at the Gustavus winter conference, ethicist Paul Ramsay of Princeton, laid out a full ethical argument *supporting* eugenics, saying many of the same things as Shockley but from a different perspective. It must be added that Ramsay's support in no way dealt with the issue of race, but rather concentrated on humanity's genetic future and what, under the standards of Christian ethics, can or should be done about it.

Ramsay began his talk with the mandatory nod to eugenics's contemptible history. 'The culmination or abuse of eugenics in the ghastly Nazi experiments would seem to be sufficient to silence forever proposals for genetic control,' he said. That should not be so for two reasons. One was that a dark, apocalyptic vision of the genetic future of

---

* Cold Spring Harbor is now an esteemed genetics research lab. Materials from the eugenics office were removed to the University of Minnesota, where they now rest, virtually untouched.

humanity – which he would describe as something close to the end of the world as foretold in *Revelations* – seemed to haunt the nights of many reputable geneticists. The other was advances in the field of genetics, which even in 1965 were beginning to clarify the issue.

Ramsay said that while geneticists disagreed on the degree or speed with which genetic degeneration was happening, they did not disagree that it was. The laws of gene frequency and the processes of mutation and selection apply no less to the 'higher' human attributes than they do to the 'lower,' he said. Mental and moral traits also have a genetic basis, he said.

The Princeton ethicist then added: 'Thus, by doing away with natural selection that used to keep us reasonably fit, by holding at bay the lethality of lethal genes and by weakening the disfavor formerly placed upon bearers of unsociable traits, mankind is allowing an insidious genetic deterioration that will leave us unfitter than we began.'

Ramsay pointed out that in Christian theology there is no imperative that humanity live forever. Indeed both Christian and Jewish theologies are founded on the notion that it won't. That does not lift from humanity the need to act; there is only the requirement that it act morally.

Ramsay said there were two ways of dealing with the situation. The first was the development of what he called 'genetic surgery,' now called genetic engineering, to eliminate the causes of genetic defects. The second was parental selection, and birth control. Whatever was done had to be voluntary. Nobody should be forced to do anything, because that was wrong and because it would have a negligible effect on the gene pool. On the other hand, if carriers with a defect could be encouraged to have half as many children as they might ordinarily, that would reduce the 'abnormal-gene frequency by 50 percent,' Ramsay said. He called it the 'ethics of genetic duty.'

'There is ample and well-established ground in Christian ethics for enlarging upon the theme of man's genetic responsibility. Having children was never regarded as a selfish prerogative. Instead, Christian teachings have always held that procreation is the place where men and women are to perform their duty to the future of the human species. If a given couple cannot be the progenitors of healthy individuals, or at least not unduly defective individuals, or if they will be the carriers of serious defects, then such a couple's "right to have children" becomes their duty not to do so, or to have fewer children.' If churches could promote celibacy as a glory to God, then 'these same Christian churches should be able to promote voluntary or "vocational" childlessness, or policies of restricted reproduction, for the sake of the children of generations to come,' Ramsay said.

The gloomiest view came from geneticist Herman Joseph Muller, a genetic Jeremiah who won the Nobel Prize in 1946 for his work on radiation-caused mutations in fruit flies. Muller, a Marxist, predicted a far future world in which 'the then existing germ cells of what where once human beings would be a lot of hopeless utterly diverse genetic monstrosities.... The job of ministering to infirmities would come to consume all the energy that society could muster. Everyone would be an invalid. The lame would tend the halt. Society would be in complete disorder. In short, humanity would end and there would be none like us to follow.' Muller himself proposed several things under the rubric 'germinal choice,' including artificial insemination of the best and the brightest, possibly from sperm banks. He turned out to be ahead of his time. It is now routine, for instance, for Ashkenazic Jews to be tested for the Tay–Sachs disease-causing mutation, and for couples of Mediterranean background to test for thalassemia.

What is important to note about Ramsay and Muller's views is that the underlying discussion of eugenics, including many of the things Shockley was talking about, was not the solitary ravings of a lone paranoid that no responsible moral person or scientist would utter. Ramsay, Muller and many others made a case that it was immoral *not* to take the sorry shape of the world's population seriously. For a number of reasons this ethical position got lost in the noise. So Shockley had considerable, if faint-hearted, support in the scientific community. The underlying concern – forget race for a moment – did not ooze up from a sewer of bigotry. The issue Shockley raised was one accepted by other respected academics and had real – if inconclusive science – behind it.

But Shockley was a terrible advocate for such a nuanced and controversial cause, letting his opponents paint him into foul corners and letting them assume the moral and scientific high ground. And he would take his arguments into places where he probably should not have gone.

CHAPTER 11
'What law of nature have you discovered?'

A few months after the Gustavus speech, Clevite gave up on Shockley Semiconductor and sold it to International Telephone & Telegraph, which moved it to Florida; when IT&T couldn't sell it, they killed it. That freed Shockley to pursue a career different from the one he had planned.

In April 1965, he contacted his friends back at Bell Labs and asked for work.[52] One executive, Jack Morton, seemed pleased to get Shockley back into the fold, but the labs really had nothing for him to do. 'The only thing I can think of at the moment is that you and I should... talk about trying to get something started on scientific problem solving for our first-year trainees – particularly during the summer before they go off to their respective universities,' Morton wrote.[53] Essentially, Shockley, the Nobel Laureate in physics, was going to bring his freshman seminar to Bell Labs. He was perfectly willing. He was sure it would be useful.

Shockley also needed the money. He drew a salary from Shockley Semiconductor, but never made a dime on the stock. Bell Labs offered him $2,000 a month for 40% time at the labs, including expenses, but not including the air fare back and forth between California and New Jersey.[54]

Stanford meanwhile offered him $39,000 for the other 60% of his time. Bell later gave him a raise, but he eventually cut back to 20% of his time there. Too much else now was happening in his life, including the federal grant to develop his program 'Mental Tools for Scientific Thinking' for the Palo Alto public schools. That grant was a big one, $100,000, and would require a quarter of Shockley's time. What he learned in that program, he assured the labs, would be useful in teaching their interns. Again, he was trying to train students how to think about and solve problems. He was even considering writing a book using the technique, he told the labs.[55]* The labs agreed and gave him sufficient time off.

By now he also was moving his Stanford time away from physics and electrical engineering, although he was still teaching an occasional semiconductor course. His schedule got so complicated that he had to draw out bar graphs to keep track of what time he owed and where.

It was clear to Shockley that he was proposing research that a great many scientists seriously did not want done for reasons that had nothing to do with science. He believed they were afraid of the results. Before the Commonwealth Club of San Francisco, a public affairs organization, in January 1967, Shockley went after his critics in a speech entitled 'City Slums and Research Taboos – A National Sickness Diagnosed.'

Shockley said America was infected with denial. '[A] wishful-thinking microbe has paralyzed our ability to doubt in the face of the desire to believe, so that contrary opinion and even proposals for research are rejected,' he said. 'In fact, recognized intellectuals use such emotional language as "a basis for repression and murder" to attach taboos to such proposals.' A loss of objectivity is the first symptom of the germ, he said. Part of the problem, Shockley agreed, was a widespread but unfounded belief in the unlimited plasticity of intelligence. In this theory, it doesn't matter what intelligence you are born with, environment and education can alter those native gifts so that all people can be made relatively equal. If that doesn't happen, it's the environment's fault. He said that this theory derived from 'inverted liberalism,' a product of which, he said, was the urban slum – mostly inhabited by African-Americans – and the inability of many recruits in the military, 'by no means restricted to Negroes,' to pass qualifying exams.

Surely environment plays a role, Shockley said. But is it the only cause? 'Is perhaps some of the cause heredity? Can anyone answer these questions? I have searched and found only unconvincing assertions that carry no sense of certainty.' Shockley said that when he raised those questions the outrage in the scientific community was stunning. The nerve being hit was the one of race, Shockley said, but as

---

* He wrote one, *Mechanics,* coauthored with Walter A. Gong. Shockley's program eventually spread to six Bay Area junior and senior high schools.

he put it there was no way you could touch the environment–heredity problem without getting to the 'Negro problem.' Shockley said the evidence was strong that there was a racial difference in intelligence, genetic in origin. More than a few scientists agreed with that, he said, some of them privately, some of them publicly, including two past presidents of the American Psychological Association. He received supporting letters (more of which later), he added, almost all from other scientists who were afraid to express their opinions in public.

Shockley said he was all in favor of every improvement in the environment for people. But he said that finding the truth in the nature–nurture debate 'would contribute to setting us free to improve the welfare of man.'[56]

The audience received the speech warmly.

The next morning, science writer David Perlman reported the speech in the *San Francisco Chronicle*, generally writing what Shockley had said and that the audience was receptive. Perlman, considered by his colleagues a dean of American science journalism, reported it straight and fair.[57]

Shockley, beginning a pattern he would continue for many years, decided he could leave no news story unanswered if there was a possibility of further publicity. He fired off a long letter to the *Chronicle* blasting Perlman's coverage of the talk. Then Perlman did a silly thing: at the suggestion of an editor, he answered back, defending himself, in a letter to his own newspaper.[58] Perlman's response gave Shockley the opportunity to get in another word, in yet another letter to the editor, which the *Chronicle* also dutifully published.[59] Shockley's response was not only longer than Perlman's original letter; it was longer than the original story. Perlman got in the last word in a note at the end of Shockley's letter, joking that Shockley was an expert in genetics, medicine, politics, sociology and now journalism. 'This, indeed, is virtuosity.' He ended the conversation and never repeated his mistake.

Shockley's love–hate relationship with the media had begun. By 1967, he had become notorious. Since the *U.S. News* stories, he had been quoted in newspapers all over the country. Stories about his theories generated news, and they invariably generated comment and letters to the editor – many from Shockley, but many from readers who found the stories provocative in the extreme. Correspondence both pro and con swamped newspapers. Often the letters blasted the newspapers for giving Shockley a forum in the first place, which, correspondents wrote, lent his 'racist' theories credibility. Others countered that the newspapers were doing what they are supposed to do. Sometimes Shockley would comment on letters commenting on the news stories. Often his rebuttals were printed.

His speeches now attracted violent demonstrations. Those demonstrations also made news, giving Shockley perhaps more publicity than the speeches alone were worth. To him, there was almost no such thing as bad publicity. He instigated coverage with staged events: debates with other experts, many of them African-Americans, that drew crowds and reporters. At first, they went peacefully, but they soon degenerated into name-calling and went downhill.

The debates took on a familiar pattern. Shockley remained impassive as those about him became emotional, perhaps one reason Shockley won sympathy and respect from some in the audience despite his terrible abilities as a debater. Opponents' insults often seemed to surprise audiences, who were not expecting to see scientists behaving badly. Many, particularly the African-Americans, were unsurprisingly offended by what Shockley was saying; some lost their cool, a failure that became self-defeating. Anyone innocently attending such a debate and wanting to hear a full accounting often ended up hearing only one side of the story – Shockley's. Everything else was an *ad hominem* attack. Audience members may not have liked Shockley very much – his stage persona was dull and cold and his message difficult – but many of them probably ended up liking his opponents even less.

He was alone in public, and at this stage his standing as an educated amateur was a major weakness in his position. He got help.

In 1968, Arthur Jensen, a professor of educational psychology at the University of California at Berkeley, was a fellow at the Center for the Advanced Study of Behavioral Sciences at Stanford. Scholars in social sciences were – and are – invited for a year of quiet to research and write on a hilltop in beauty and solitude, interrupted only occasionally by a stray chip shot from the golf course below.

The center invited Shockley to talk to the fellows about his 'human quality' projects and Jensen, who was working in the field, went to the lecture. Jensen said later that he was impressed. He was researching intelligence tests, especially one that Lewis Terman used on his 'gifted' subjects, and he wanted to get as many bright people as possible to take the test. Jensen asked Shockley, who agreed. Shockley scored only in the ninetieth percentile. The test was too verbal for him, Jensen says.

They chatted several times and Shockley invited Jensen to dinner one night. Jensen had never met a Nobel Laureate before, let alone had dinner at home with one, and eagerly accepted. Shockley wanted to talk about Jensen's research and his interest in intelligence, and sent Jensen some material to read beforehand. Jensen was busy and didn't have time to do more than look at it. Shockley began quizzing him over dinner about one of the papers he sent. Shockley was convinced the author had misused a statistical test called chi-squared. Jensen had to admit he had not read the paper very carefully.

'Is that how you people in behavioral science do your homework?' Shockley said. 'No wonder you are in such a mess. I have better things to do than talk to you!' He got up from the table and went into his study to work, leaving Jensen and Emmy staring at each other.

'Don't worry Art,' Emmy said. 'He would do that to the president of Stanford.'

Next morning, Shockley called Jensen to suggest that maybe next time the psychologist would do his reading so they could have an intelligent conversation. Jensen went back to the Shockleys a few nights later – properly prepared.[62]

Shockley by now was becoming a churl. Jensen tells of another incident in which he, his wife, the Shockleys and a friend from Berkeley and his wife were having dinner. The friend, described by Jensen as a distinguished social scientist, had expressed an interest in meeting Shockley. During dinner, the friend had the temerity to contradict something Shockley had said, a minor point. With fire in his eyes, Shockley looked at him and said: 'What field did you say you were in? What law of nature have you discovered?'[62]

Jensen stayed on as a friend despite the need for a certain insensitivity. For one thing, he was taken with Shockley's brilliance, especially in statistics, which Jensen found useful. He could bounce statistical matters off him. He remembers one occasion when he showed Shockley an article from *Science* which apparently contained a subtle but fatal mathematical error. Jensen said it was a good test for graduate students to spot the mistake. The journal editors surely missed it, as did the scientists who refereed the article. Statisticians at Berkeley also struggled. Jensen himself took a half hour to find it, although it was a short paper. He showed it to Shockley, who read it through in a few seconds and then announced, 'The guy did this wrong.' He then provided the correct answer.

In the winter of 1969, Jensen dropped his bomb. Writing in the *Harvard Educational Review*, Jensen changed the nature of the debate in an article entitled 'How Much Can We Boost IQ and Scholastic Achievement.' Jensen had been asked to write an article on why many of the social programs of the War of Poverty seemed to be failures. He wrote that it was 'not unreasonable in view of the fact that intelligence variation has a large genetic component, to hypothesize that genetic factors may play a part' in the poor performance of many disadvantaged children in school. 'But such a hypothesis,' he wrote, 'is anathema to many social scientists. The idea that the lower average intelligence and scholastic performance of Negroes could involve not only environmental but also genetic factors had indeed been strongly denounced. But it has been neither contradicted nor discredited by evidence.'[63]

Jensen made four arguments:

- Compensatory education for disadvantaged students has failed to raise their IQ scores.
- Children with low IQ scores are both genetically and environmentally challenged, so efforts to raise their IQ through just education are doomed.
- Genes play a major role not just in differences among individuals within groups, but also in the IQs between groups.
- The way to handle these children in school is rote learning, not the teaching of abstractions they cannot grasp.

Until then, it had been easy to denounce the idea of a genetic disadvantage in African-Americans as the ravings of a physicist stumbling out of his field. Jensen stopped that argument dead. This was his field. He was nationally renowned for his research, was tenured at Berkeley and directed the Institute of Human Learning there. Within days of publication, national magazines and syndicated columnists picked up the story. Within days of the publicity, Jensen became a pariah, attacked and vilified by almost everyone in the scientific establishment. Whole scientific papers were dedicated just to discredit 'Jensenism,' which often became a noun with a small *j*. One critic, Jerry Hirsch of the University of Illinois, could rarely refer to the paper without putting the word 'notorious' in front of Jensen's name.

Jensen's life changed dramatically. His office was picketed and students demanded he be fired. The threats became so serious the campus police assigned him plainclothes bodyguards.

The journal itself was astounded by the reaction, all of it negative. How dare they publish such a thing! The editors published rebuttals by seven authorities, and when that didn't quell the rumpus, published some more. Then the editors claimed they never asked Jensen to write about race, but he produced the solicitation letter from the magazine. So the editors just stopped selling that issue, even for reprints, even to Jensen.[64]

Here was Shockley's scientific cover. He had at least one prominent specialist who agreed with him and was brave enough to stand up. In some ways, this took the pressure off him. Jensen became the main target for those who thought the differences between people were largely environmental.

Some scholars, including the liberal psychologist Christopher Jenks, called the debate a draw. Jenks wrote that Jensen's argument wasn't all that persuasive, but that the critics could offer no persuasive evidence that he was wrong.

By May 1969, the opposition had united, including the Society for the Psychological Study of Social Issues, made up of 18 prominent psychologists. They essentially agreed with the National Academy of Sciences that no one could measure the effects of heredity on African-Americans, especially when African-Americans were suffering from social inequalities. One member pointed out that attorneys fighting integration in public schools in Virginia had quoted Jensen's article.

The scientists' points were:

- There is no 'direct evidence' that there is an inherited difference between blacks and whites, though, they agreed, African-Americans scored regularly below whites in IQ tests.
- Racism and discrimination impose 'an immeasurable burden' on African-Americans and prevent them from living comparable lives to whites of the same social class.
- Environment affects a child's development from the moment of conception.
- Present IQ tests are culturally biased against African-Americans.

The society also mentioned that identical twins raised in different environments can 'show differences in intelligence test scores fully comparable to differences found between racial groups.' They would regret mentioning the twin studies.

This was the kind of debate the topic merited. There was no name-calling or intemperate attacks. Scientists differed on the issues, not on the nitty-gritty of arcane statistical gymnastics or on motivations. Unfortunately, this kind of reasonable discourse would be rare.

The debate took place against the backdrop of the great student rebellion in the late 1960s, triggered by the Vietnam War. The issue was not just the war, however, but the revolt came with all the passion of the civil rights movement. President Lyndon Johnson signed the Civil Rights Act in 1964, and marching around the country proclaiming that African-Americans were intellectually inferior to whites was at best insensitive. Shockley was tone deaf to the times. This set the scene for violent demonstrations and political extremism. Shockley had some respite from the protests at home because Stanford's laid-back students were famously late joining the uprising. When the student revolution finally hit the Stanford campus, it did so with a vehemence that surprised even the students. Soon campus buildings had ground floor windows boarded up permanently: the administration got tired of replacing them. Some structures built in that era were especially designed with as few windows as possible, especially near the ground. To this day, they stand out in the normally open and airy campus. This was no time to raise controversial issues about race and equality if you wanted a quiet, rational debate.

Shockley was concerned that the Stanford administration under then-president Kenneth Pitzer was retreating under pressure from the student rebels. He spent some time with a leader, a graduate student in physics, and reported his impressions to Pitzer, warning him not to give in. Pitzer, who gravitated to the path of least resistance, ignored the advice.[65]

Shockley's opponents were clearly limiting his right to speak. An appearance at the Brooklyn Polytechnic Institute in May 1968 was cancelled for fear of violence. Shockley and almost 500 other scientists were invited to a symposium entitled 'What Can Man Be' to honor the 50th anniversary of Sigma Xi, the academic science society. Those who invited Shockley knew perfectly well what he was going to talk about. A small group of faculty at the institute responded to his announced appearance, calling the Sigma Xi planning committee Nazis, racists and representatives of a lunatic fringe. The committee asked Shockley to talk about something else. He refused. Sigma Xi immediately canceled the whole convocation, despite the fact 493 scientists and engineers had agreed to come. The society sent telegrams announcing the cancelation due to 'serious developments.' Nobel Laureate I. I. Rabi thought the cancelation was a good thing. 'At the moment we ought not risk demonstrations because we do not have enough police.'

Not everyone agreed. Harold Taylor, former president of Sarah Lawrence College suggested, 'If we are never to discuss any controversial issues for fear there might be demonstrations then the whole purpose of the university's destroyed.'[66]

The good news for Shockley was the beginning of a backlash. Often, after he was prevented from speaking, conservative or civil liberties groups invited him to speak as a form of protest, and newspaper editorialists would rally to his defense. Shockley found that his strongest allies, then and later, were editorial writers, some of them at liberal newspapers, who were aghast at the implications of censorship by physical threat and by pusillanimous administrators and bureaucrats who caved in to the threats. If anyone was behaving like fascists, the editorial writers said, it wasn't the people who invited Shockley.[67]

Shockley had a more immediate problem at Stanford, however: the university had hired him to teach physics, not genetics. He decided he was going to put off all his other work and concentrate on this issue. 'I am dropping certain activities in the physical sciences with the intention

Figure 20 Shockley teaching his electronics class at Stanford in the 1970s.

of attempting to determine on my own whether intelligence is equally distributed among all races of man,' he said.[68] To devote the time he needed for his new passion, he needed to raise money himself. If he could present the university with funding for his projects, the school would have little choice but to let him continue. He relied on the administration's sense of academic freedom and inquiry. He sent out form fund-raising letters to people who had expressed an interest in his research and the newly named Human Hereditary Quality Fund.

Shockley said the fund would help him set up 'sound methodology' to research human quality problems. 'Emphasis will stress activities for which benefits to mankind will occur in the next one or two generations.' He said he could make no commitment about results and, except for obeying the rules of the university, he would exercise complete discretion in the use of the funds. He wrote that donors should contact the vice provost, Herbert Packer.[69]

He got enough of a response (tens of thousands of dollars) to satisfy the university. One response to his appeals came from a New York lawyer, Harry F. Weyher (pronounced wire). Weyher happened to be on the board of a foundation called the Pioneer Fund, and the publicity over the Brooklyn Tech cancelation convinced Weyher that Shockley was doing work the fund might support.

A reclusive Massachusetts textile-machine magnate, Wycliffe P. Draper, Harvard class of 1913, founded the fund in 1937. Draper, a eugenicist, was interested in funding research that would support his theories that Negroes were inferior. Draper received some notoriety in 1960 when some scientists refused to take his money to do studies of that nature. Others were not as reluctant. One founding director was Cold Spring Harbor's Harry Laughlin.

Weyher ran the fund from his Fifth Avenue law offices on a volunteer basis. No staff, no offices. The fund's resources were not vast. When Draper died in 1972, he left about $1.4 million to it, and at the time Shockley entered the picture investments had brought it up to about $5 million.

The Pioneer Fund became part of the controversy. Its decidedly right-wing slant, and the fact that some certified racists were funded by it or supported its work, did to the Pioneer Fund what the Nazis did to eugenics: mention the fund and unsavory associations were inevitable.[70,71] Yet the fund sponsored some of Jensen's research in future years, as well as the University of Minnesota twin studies, the largest and most respected of the type.

With funding in hand (how much is not known), Shockley proposed to Stanford that he increase his load at the university and begin a human quality program. He was told to submit a formal budget.[72] In May 1968, he asked the university to take him on full-time. Weyher, acting as an agent for several clients, including the inventor William Lear, told Stanford he had some funds available for Shockley, and Shockley told the university to use the Lear money for the increase in his salary. Shockley also asked for $4,600 for office expenses, the cost of three transcontinental trips and a 'lunch conference.'[73]

Pettit, the dean of engineering, was willing, but needed a better understanding of what kind of research Shockley was proposing. It did not sound like physics. He also needed to know how Shockley was qualified to do such research. The subject was controversial and certainly beyond the scope of Shockley's proposed activities when Stanford hired him.[74] Vice provost Herbert Packer assured Shockley the university had no intention of censoring its faculty. Shockley asked that the money be placed in a discretionary fund that he could use as he wished, but Packer said that was against university policy. He needed a real budget so that everyone was sure where the money was going. Pettit also wanted to make sure the money didn't come from the engineering school's pot.

Pettit passed Shockley's proposal to several other faculty members, all of whom had serious qualms. He wrote an office note to another administrator on the results of their inquiry, which found its way to Shockley. One commentator, probably a geneticist, said the project was too difficult to handle 'in a research sense.' Another said it was beyond Shockley's competence and this was no area for an 'amateur to dabble in an undertaking of such sensitivity.' It might even be dangerous. A third said the research had the potential to make a bad racial and social situation worse. Pettit also noted in a memo that one of his daughters suggested that Shockley 'should be put away.' 'I don't accept this as the final word,' Pettit noted parenthetically.

Pettit said the first objection, that the research would be too difficult, did not impress him. Since when should a scientist back away from a project because of the difficulty? The second objection, Shockley's competence, was more to the point. Here, he said, Shockley was being subtle. He was not proposing actually doing the research; he was proposing studying ways to do the research, the methodology. That he was competent to do, given his reputation with operations research and statistics. It would be hard to argue that a fresh look into the research difficulties by a new mind would be 'misplaced.' He did not respond to the danger, only repeated in the note, that whatever Shockley was proposing it surely wasn't engineering.[75]

In June 1969, Shockley wrote a letter to Weyher asking for more money. The fund gave him $20,000. His proposal was muddied by the fact that he did not know just how much time he would have to spend at Bell Labs. By December, he went with a formal proposal and a budget. That letter gives a unique view of his finances at the time. Because of his teaching load and a project to revise *Electrons and Holes*, he had one day a week free to work on eugenics. He called his project 'Research On Methodology To Reduce The Environment-Heredity Uncertainty, Including Ethnic And Racial Aspects' – right up the Pioneer Fund's alley. He would start on 1 July 1970 and finish on 1 February 1975. Shockley explained to Weyher that he wanted to put more time into the project, but he had serious pension considerations at Bell Labs. He would be 60 years old the following February and 65 in 1975. If he backed away from committed time at the labs, the loss to his pension time would cost him dearly.

He did not get all the funds he needed, in part because the Pioneer Fund didn't have that kind of cash. There is no record of how much he did get. He would get tens of thousands of dollars in other donations, including one from Draper himself, but eventually he had to subsidize his own research, and he did not retire until 1975.

On the political front, he felt it imperative to get the National Academy involved, if for no other reason than the publicity.

Shockley still hoped that his relationship with Seitz would help him get the National Academy's attention, but Seitz was leaving the academy post to become president of Rockefeller University. The next year the academy meeting was in Washington and Shockley again tried to deliver a paper on the 'Cooperative Correlation,' using the economic data and the Terman study to show that something besides discrimination was responsible for African-Americans being on the bottom of the economic scale.[76]

'I took him home to spend the night,' Seitz remembered years later, 'and the only reason he visited me is that he wanted me to allow him to give a lecture on his racial theories. I just said, you know Bill, you'd blow [up] the place. This did not stop him at the open meetings from getting up and speaking.... At that stage you could not reason with him.'[77]

The new NAS president, Philip Handler, formed a committee headed by Kingsley Davis, a Berkeley sociologist, to assess Shockley's proposal. The Davis committee repeated the view that the study of racial differences was 'proper and socially relevant.' But, it added: 'In the first place, if the traits are at all complex, the results of such research are almost certain to be inconclusive. In the second place, it is not clear that major social decisions depend on such information; we would hope that persons would be considered as individuals and not as members of groups.'

The committee made three recommendations: it called for closer cooperation between researchers with expertise on the relation of heredity and behavior; it recommended that the National Science Foundation consult on the educational implications of behavioral genetics; and finally that the academy establish a working group to study the feasibility of doing long-term research.

When he first learned of the committee report, Shockley thought he had won his first victory, calling it 'an enormous stride.' Then, several days later, the academy membership voted: they passed the first recommendation, but had no interest in agreeing to the last two.[78–80] His victory was short-lived.

Shockley's motion to fund research on dysgenics was never seconded.

That meeting also saw the first attempt to muzzle Shockley by changing academy rules. Joel C. Hildebrand, a professor emeritus of chemistry at Berkeley, moved to declare it out of order for any member to ask the academy to fund one of his own research projects, which is what Shockley was doing. The Academy tabled his motion. Afterward, Hildebrand said that the Davis report was 'worthless for the purpose of making clear to the public that Shockley's proposals are essentially unscientific and antisocial.'[57] Later the Academy would change the meeting rules to make it impossible for another Shockley to drag them into another mire.

Why wouldn't Seitz support Shockley's motions? 'He [Shockley] would have liked to make this [limits on science] a very major theme of the Academy and that would not have been popular with the membership. I think he [also] would have liked to have convinced me that he was right in detail and endorsed his views about race.' Seitz, in fact, did not agree with Shockley.[77]

Shockley called the Davis committee report faint progress.

He felt betrayed by Seitz and broke off all personal contact. His criticism of his oldest friend was vicious. He called Seitz's failure to do what he wanted a low point in national scientific leadership. 'President Seitz's views... appear... to be frighteningly subservient to a popular majority opinion rather than to one tested by adequate study and debated and thus... not appropriate to a position of leadership in science,' he said.[81] He equated the defeat in the academy to a version of Lysenkoism, referring to the Soviet agronomist Trofim Denisovich Lysenko, who decided one day that Mendelian genetics was all wrong and that species evolved by acquiring characteristics. Since Lysenko was head of the Institute of Genetics of the Soviet Academy of Sciences, his erroneous view was the enforced paradigm until Stalin died and something resembling science took over. Soviet biology never recovered.

Shockley and Fred Seitz, the man he had known since the summer drive to MIT in the battered convertible in 1932, his closest friend through the war years and at Lake George, his oldest friend in the world, never spoke to each other again. When Shockley was dying, Seitz sent a note. Shockley never answered.[77]

CHAPTER 12
'Someday we may actually be terribly alone'

The best scientific support for Shockley at the time was about a thousand feet across the grass and down a slight slope from his McCullough building office on the top floor of Stanford's psychology building: the Terman study, one of the longest-running longitudinal studies in the history of science.

It's still going. A few subjects are still alive and Terman's successors now are looking at the third generation to see if what they found in the first generation can be found down the generational line. Are really smart people more likely to have really smart children? The study is one of the great icons of social science. Shockley knew about it, but how much of Lewis Terman's work he read or how critically he read it is not clear. Some information he probably received from Fred Terman, Lewis's son and the man who hired him at Stanford. Shockley mentioned the study often, but not always correctly.*

In fact, Terman's study of the gifted provides serious evidence of the inheritability of IQ, but provides absolutely no information for those looking for racial differences.

Terman, professor of psychology at Stanford, began the study in 1921. His goal was simple: He was a eugenicist (a position he softened somewhat as he got older) and he believed the future of America lay with its most intellectually gifted children. He believed society had a view of genius that was wrong and dangerous: that very bright people were sicklier, physically smaller, more emotional, more prone to mental problems,

---

* The Terman files are kept confidential because of privacy promises made by Terman to his subjects. Even they cannot see them, so certainly Shockley never did. Nonetheless, Terman published extensively. In 1990, I became the only journalist ever permitted access to the files. Little, Brown published the results in the book *Terman's Kids*.

anti-social and difficult. They start as bright, if brittle, children and end as dwarfed, repressed, unstable adults. He called this view 'early ripe, early rot.' He wanted to prove that was incorrect and to start people thinking about programs in schools to stimulate genius, coddle it and nurture it for the greater good. Every program for the gifted in America's schools to this day derives from Terman's work, and he did show that none of the myths about genius are true.

Terman began using the word 'genius' to describe his subjects, but later just called them gifted, the term we use now. His definition was simple: a genius was anybody who scored 140 or better on the Stanford–Binet IQ test.

In 1912, with the help of a graduate student, H. G. Childs, Terman revised Alfred Binet's intelligence test into a series of 90 tests (six for each age group) to measure the IQ of children. He called it the Stanford–Binet test, and for 50 years it was the standard, used by schools, companies, the military, and government. Terman was the first to make these kinds of measurements, essentially becoming the founder of psychological testing in America. Constant royalties from the test and its revisions supported his research through his lifetime.

He put together a cohort of children who scored the highest on his tests. He set the bar first at 140, but when he failed to collect enough students who could meet that standard, he lowered it to 135.

The tests were measuring an ambiguous human attribute or portmanteau of human attributes. Scientists need numbers to make comparisons or measure change, and that is the true purpose of the Stanford–Binet test and those that followed – it gave them numbers to work with. Exactly what those numbers meant – mean – is the big question. Until very recently, psychologists defined IQ as whatever IQ tests measured, a circular justification of no help to anyone. The tests are well-known for being good predictors of how children do in school, but not much else. They measure something, but what?

Terman eventually had around 1,500 California kids in his database. 'Termites' knew they were special. If they forgot it, Terman himself was around with another test to remind them. He intervened in their lives to give them a hand (money, college entrance, job recommendations), one of many flaws in the survey. Terman measured them against groups of children with lower IQs gathered by other researchers, particularly scientists at Berkeley, to see differences. The ungifted acted as something like a control group. He compared Termites to each other – how the ones who succeeded in life differed from those who did not. The data grew to fill a small room with filing cabinets five feet high.

Terman reflected the classist attitudes of his time. His graduate students prowled schools mostly in the upper-class and upper-middle-class neighborhoods of San Francisco and Los Angeles looking for subjects, because Terman was sure that was where he would find them. City slums, he felt, would be a waste of time and resources. He felt strongly that whites were smarter than blacks, so when he went looking for geniuses in California schools he stayed away from schools with a large African-American attendance. In fact, only two black children were in the study, and both of them, he noted, were of mixed blood. His sample was seriously skewed in other ways as well, with too many boys relative to the girls,* far too many Jews to gentiles with respect to the general population, not enough Asian-Americans and virtually no Hispanics or children of Mediterranean origin. Mostly it was because of where he looked. As time went on, he became aware of some of the deficiencies and tried to address them, with only moderate success.

The main result of these inadequacies is that specific examples of differences between his gifted children and the rest of the population need to be taken with a high degree of skepticism; his study of the group itself, however, stands unchallenged.

Termites were slightly larger, healthier and better adjusted than children in his control group. Some of that surely was the fact that they came from prosperous, stable families who could afford the best medical care, food and housing. Whatever the reason, they were very bright and they were not physically or emotionally handicapped. IQs changed very little through the years, no matter what happened to their environment. When IQ did change, it seemed to go down. Terman didn't know why, but it was likely an inconsistency in the sequence of tests.

The gifted children generally came from gifted – or at least successful – parents just like them. Thirty per cent of their fathers were professionals; 46% were in business; 20% were in industry. Only one was a laborer.

The Termites tended to marry bright people, usually almost but not quite as bright as themselves.

---

* No one knows why there is a gender gap. Teachers, mostly female, recommended most of the children and they may have recommended a disproportionate number of boys to girls. The reason may have been a cultural bias in the tests, or perhaps boys just do better at these tests.

Terman's kids were wonderful students. Eighty-five per cent skipped grades at least once. Three-quarters of all their grades were As. Fifty per cent could read before they got to school. When they reached college – and they did in unusually high numbers – they practically gobbled degrees. A third of the Termites admitted to Stanford graduated Phi Beta Kappa. Terman found nothing in their home lives (and his researchers inspected many of the homes) to suggest the influence of environment to explain this success, he reported.

IQ also is a good predictor of success in life, at least by conventional middle-class standards, and the Termites did splendidly, becoming (at least for the males) doctors, lawyers, businessmen, and scientists at a vastly higher rate than would be expected from the general population. The 1,500 children grew up to produce at least 2,500 scientific articles and papers, 200 books, more than 400 short stories and 350 patents. And that didn't count the output of the professional journalists. Terman was so proud of them that his files bulge with their work. Three were members of the National Academy (including his son, Fred); six made the *International Who's Who*; 40 made *Who's Who in America,* and 81 (including 12 women) made *American Men of Science.* Terman's kids worked for the Federal Reserve, the Atomic Energy Commission, the staff of the US Senate, the Department of Justice, NASA and the United Nations. During the Second World War, the men earned 90 valor medals, including 15 Purple Hearts. By and large, they reported themselves to be happy people, and they lived longer than the population average.

The most important findings for this discussion concern the Termites' children. They also were exceptional, with a mean IQ of 133, slightly less than their parents but still around the 98th percentile.* Sixteen per cent of the children of the gifted were themselves gifted, an astonishing percentage – in the general population you would expect less than 1%. Less than 20% of the second generation had IQs below 120, also an amazing figure. Statistically, that is impossible to explain away.

Terman was convinced he was watching heredity.

---

* The decrease is probably what statisticians call a regression to the mean, and is expected in such a study. More about that later.

Most of the criticism of Terman's results points out that the children (and their children) started life with all the advantages of prosperity and continued in that kind of atmosphere as they grew up. But Terman included environment in his analysis, and the premise is not entirely true to start with – a substantial number of Termites were not so financially advantaged, despite where Terman found them. Environment seemed to make no difference in their scores. It seemed to play the greatest role in his analysis of why some of the gifted succeeded greatly in life while others did not. It was unrelated to IQ.

A more valid criticism – and this is crucial to understanding the flaw in Shockley's argument too – lies in what IQ did *not* measure in Terman's study.

Most obviously, Terman missed the two Nobel Laureates. Neither Shockley nor Luis Alvarez had IQs above 135. Shockley was tested twice and missed both times. Whatever talent they had went unmeasured by Terman's questions. One hypothesis is that the tests do not measure mathematical prowess very well, but is that ability not a facet of what we mean by 'intelligence'?*

One of the great mysteries of Shockley's story remains: how could someone who was a living embodiment of the weakness of IQ tests destroy his reputation on a theory based on their credibility?

Part of the answer may be merely reading the results with an astigmatism, a lens bent to show what you want it to show; part of it could be simply that neither Shockley nor anyone outside the Stanford psych department – with few exceptions – had seen the files, so they actually didn't know all that was in them. For instance, Shockley often said the Termites won an uncommon number of Nobel and Pulitzer Prizes. In fact, none of Terman's kids won either prize. No one even came close.

Another gap is the issue of creativity, a deficiency that bothered Terman so much that he sent his investigators to a school for the arts in Los Angeles to test children, hoping to include that variable. They all flunked. Terman found no link between IQ and creativity, particularly in

---

* There have been five sets of Nobel parents and offspring: Niels Bohr (1922) and his son, Aage (1975); Ulf von Euler (1970) and his father Hans von Euler-Chelpin (1929); G. P. Thompson (1937) and his father, J. J. Thompson (1906); Irene Joliot-Curie (with her husband in 1935) and her mother, Marie (1911 and with her husband 1903) and the father–son team of H. H. Bragg in 1915).

the arts. Except for Henry Cowell, who was in a different study, none of the gifted subjects became known as composers, musicians or artists, and there were only a very few writers of any note: the science fiction author L. Sprague de Camp and William A. P. White, who used the *nom de plume* Anthony Boucher. Only one, the actor Dennis O'Keefe, had any reputation in the theater or films, and another, Shelley Mydans, was one of the few who earned note in journalism. The most famous Termite was Jess Oppenheimer, the comedy writer who created 'I Love Lucy.'

Whatever strengths IQ tests measure, artistic creativity certainly isn't one of them.

If IQ can be inherited, artistic creativity seems to pop up at random. Except for Vienna's Strauss family there are few multigenerations of great composers. Mozart's father was a musician of – to be kind – modest talents, yet he sired perhaps the greatest musical genius of all, and the family hasn't been heard from since. Bach, the son of a musician, produced three sons of some musical note, but none of them could match the old man in genius. Dickens was the son of an undistinguished office clerk and had ten children of no unusual talent. Indeed, there are only a handful of great dynasties of writers (the Huxleys, the Amises), and except perhaps for the Wyeths and Brueghels, few dynasties of great painters in the history of art.*

None of Terman's kids was a professional athlete. No one founded an industry. And no one became a Mother Theresa, a cardinal or a chief rabbi. All these accomplishments can lead to a happy, successful, rewarding life, all can lead to financial reward and power, and not all of the attributes that lead you on those paths apparently are measured by IQ tests.

Neither are such things as determination and attitude.

And, it must be noted again, that because of the bias in Terman's sample, the study says nothing about race or ethnicity.

The debate should have centered on what those tests did measure and whether those things really are important to know.

---

* Acting seems exempt from this rule. Witness the Barrymores and the Redgraves, to mention a couple.

The human quality problem and the controversy surrounding it became a full-time obsession for Shockley. It released dark forces in him that seemed always to have been lurking. Only by the greatest effort – an effort he would rarely expend – could he talk about or think about anything else. Even when he visited Alison in Washington, that was the topic of dinner conversation. Alison and her husband learned to just sit and listen.

Shockley was oblivious to what he was doing to himself. There should have been an apprehension at one point that he had gone too far with his racial theories, a recognition that he had left a crucial opening for his opponents. Yet there is no indication of it in his papers and Emmy was quite clear he harbored no doubts, either about what he was saying or the wisdom in saying it. He was impelled forward by his own demons.

In the early 1970s, a Stanford psychiatrist told a reporter he thought Shockley was suffering from the classic symptoms of paranoia.[46] Indeed, he began demonstrating many of the symptoms of what is now called Paranoid Personality Disorder. Others have speculated that Shockley was a high-functioning autistic or had Asperger's Syndrome, or that he had obsessive compulsive disorder. We'll never know.

Shockley had AT&T install recording devices on all his telephones – home and office. Every conversation was recorded, and every one was interrupted by a beeping sound every ten seconds. Conversations on each cassette were separated by a countdown. Writer Rae Goodell, who researched her PhD dissertation and subsequent book partly on Shockley, recalls one conversation beginning: 'Goodell three, Goodell two, Goodell one, Goodell zero. The time is now ten minutes to seven on Tuesday, the twentieth. Goodell zero.'[46] Jensen says he knows several people who refused to talk to Shockley on the telephone because of the recordings. They instructed their secretaries that if Shockley called, they were not in. They objected less to talking to Shockley than to the fact that he was taping their conversations.[62] Some were scientists who agreed with his position but wanted to remain anonymous. He would not turn it off. He felt it was his only protection against people later denying they said something.[82]

He and Emmy, who had by this time become his assistant, entered every telephone conversation into a logbook in chronological order, with time, date, party on the phone and a brief description of the call. Every call, no matter how minor, was taped, indexed and stored. If the Shockleys ordered take-out, the conversation is likely to be on tape in his archives. If for some reason the tape recorder didn't work, he would have someone, usually Emmy, listen on the line and take dictation.

They did this for every telephone conversation for the rest of Shockley's life.

He said he thought the tape recorder was the single most important application of the transistor.[46] They made, he said, 'a profound difference in honesty.'[83] He also taped every personal conversation, usually – but not always – with the knowledge and consent of the person he was speaking to. He could prove Henry Kissinger's aphorism that even paranoids have enemies.

In 1973, Shockley was told he would be given an honorary doctorate at Leeds University in England. When he got to London on his next trip, Lord Boyle, the university's vice-chancellor, invited him for drinks. Shockley was being given the award for his transistor work, and only after the decision was made did the administration at Leeds learn about his new interests and his notoriety.

In the splendid bar at the Carlton Club, Shockley put a tiny portable tape recorder on his lap under the table. Lord Boyle didn't know he was being taped until later in the conversation, when Shockley reached down to turn over the tape. Meanwhile, the conversation took on an interesting tone. Shockley said later that he detected a certain troubling timbre, one that led him to ask two questions: 'Are you trying to lead up, Lord Boyle, to saying that you would like me to act in a particular way in respect of coming to Leeds, or are you leading up to saying you would like me to forget the offer of a degree?'

Lord Boyle said they would like Shockley to forget he had been offered a degree.[46,60]

Shockley immediately called a newspaper reporter he knew back in the US for advice on how to get some publicity and played him the tape. He was scheduled for an interview with a *Times* reporter the next day and the advice given was to mention it during the interview. The American reporter then called the Associated Press to make sure the interview was covered, thus ensuring maximum attention. Boyle and Leeds got the publicity they deserved. LEEDS SNUB FOR NOBEL SCIENTIST, read the headline.[46]

Shockley took the same care with correspondence: every piece of mail, from the Shockleys or to them, was saved in a color-coded folder. Every piece. Every Christmas card, every discussion over a bill with the television cable company, every letter, every notice of a magazine subscription lapsing. Every receipt from Federal Express for packages coming or going was recorded. Every piece of paper was indexed and cataloged. The files still contain a note that Emmy sent to General Foods about a Jello recipe. Shockley had a clipping service and virtually every newspaper and magazine clip about him or about his obsession was indexed and saved.

'I saved everything because Bill said everything that goes into the office goes into the files, and I did that,' Emmy explained. 'We were not like ordinary people.'[84]

His secretary at the time was Mary Clouthier, who had worked for Gibbons and Linvill. Her main task was to fill requests from people asking for more information on Shockley's research. She worked out of the office from the time she started in the mid-1960s, and worked at home for long periods to tend her young children, mailing packages from there. She ended up with four five-drawer filing cabinets at home. Shockley and Emmy numbered every document in his research, and pre-pared a stock packet for inquiries. Shockley would call Clouthier and tell her to either mail a packet or to send document number 476 to some-one. The Shockleys apparently had the pertinent numbers memorized.

Shockley would sometimes send material to Clouthier's house in a cab for typing. If he needed something mailed out immediately, which he did most of the time, she would put it in envelopes, stack it in a box and drive to Shockley's home. He and Emmy would be up waiting.

When her second son went to school in 1972, Clouthier returned to the McCullough building office, mostly, she said, to help Emmy. Emmy worked all day in the office while Shockley mainly worked at home. Clouthier said Emmy could not leave until Shockley called and told her she could come and make his dinner, sometimes as late as 10 o'clock.

'We had shelves of reprints,' Clouthier said. 'We were cutting and pasting... article after article.' It was, she thought later, much like kinder-garten. 'We logged every bit of correspondence that came in and every bit of correspondence that went out: CO for outgoing and CI for incom-ing. Everything was given a number and then we'd put it on a database and at the end of the month, we'd print it out in alphabetical order. I don't know that he ever used it for anything.' Clouthier said she and Emmy used to laugh that if there was ever an earthquake, they would likely be buried in Shockley's collapsing files and they would have to be dug out.

He was not an easy man to work for in many regards. 'The thing with Shockley was that you never got finished. He would have an idea of how he wanted something done, and you would start doing it that way, and you would set up this whole system and you would spend hours and hours into this. Then he would decide he didn't want it that way, he wanted it another way. Nothing ever got finished.' He was becoming disorganized.

Shockley had two offices, even when he retired in 1975. Emeritus professors usually only get one, but he knew how to play the political game. Most in-fighting at universities is over space, not position or power, and he was good at it.

One day, after an incident of typical Shockley boorishness – she wouldn't say what – Clouthier quit. Emmy felt personally betrayed, Clouthier said. When she went back to work for Jim Gibbons, whom Emmy did not like because she thought he once tried to take away one of Shockley's offices, she stopped speaking to Clouthier, for years cutting her dead when they met in the hallway.

So, the burden of being his secretary and his assistant fell on his wife. Emmy, the woman he married because of her sharp intelligence and independence, had long since suspended her own life to serve him. 'Dr. Shockley can get involved in something called mankind,' she said. 'I can't... It overwhelms me to think about mankind... I'm trying to help him do what he wants to do, and my relationship with him is important, and when it gets not to be important any longer, then I'll have to do something else about this.'[46]

She had totally and willingly melded her life into his. She became not only his wife and lover, but also his secretary and assistant, sounding board and defender, organizer and factotum. They virtually merged into one identity. She loved him deeply; he depended on her completely.

His obsession had another effect: it eventually drove away most of their friends. His paranoia and insensitivity made spending time with him more unpleasant than most people thought he was worth. Add to that the attacks, in some cases from old friends, and life for the Shockleys became a lonely stand in a walled castle.

'Someday,' Emmy told Rae Goodell, 'we may actually be terribly alone.'

―≺

The protests by students and those offended by his theories finally reached physically into his Stanford life. On 18 January 1972, 15 young people, most of them non-white, some, if not all of them Stanford

Figure 21 The crowd fills the auditorium for the 1973 race–intelligence debate.

students, invaded Shockley's class room at 127 McCullough, where he was giving a quiz on electrical engineering.[85]

On 16 February 1972, a group of students burned Shockley in effigy and vandalized his car. A group of black students, with the help of some faculty members, demanded that the Academic Council at Stanford 'undertake an examination of the role and activities of William Shockley.'[85,86] The council refused. The students demanded Shockley debate his position with three Stanford faculty, two of them black.

The great debate was held on 3 January 1973, with psychology professors Cedric X (Cedric Clark) and Dubois McGee, and Luigi Cavalli-Sforza, one of the world's finest geneticists, who would become one of Shockley's calmest and steadiest critics.

During the debate, Shockley expanded on one of his more extreme and unfounded proposals: 'For each 1 percent of Caucasian ancestry, the average IQ of American black populations goes up approximately one IQ point.' He recommended that Stanford students supply the blood for an experiment to see if that were true. Cavalli-Sforza suggested that Stanford students hardly qualified as average, black or white.

Cavalli-Sforza said he was 'very embarrassed at having to note that some of the main concepts of genetics have escaped Professor Shockley. Just looking at the results of tests says really nothing about genetic differences.' If Shockley were his student, he implied, he'd flunk him. The

Figure 22 Shockley at the debate, Luigi Cavalli-Sforza on his left.

same thing would happen to him, he admitted gently, in Shockley's physics class.[87] This debate, in front of a huge crowd, went without incident. The other two said nothing memorable.

Shockley and Cavalli-Sforza did it again at a synagogue in San Jose, again without incident.

At about the same time, Shockley was invited to submit a proposal for a course with a unique Stanford program called the Stanford Workshops on Political and Social Issues (SWOPSI). The program essentially was

Figure 23 Shockley fails to get his point across.

created for courses that fell through the cracks of the academic structure or didn't radiate enough prestige to be seriously considered by a department.[88] Shockley submitted a proposal for the Winter Quarter 1973.

The SWOPSI board couldn't reach a decision. Dan Lewis, the SWOPSI director, told Shockley they would try again for Fall Quarter 1973.[89] The reasons for the stalemate were the same as the faculty committee's: Shockley's limited qualifications for the course and the fear that the course would do more harm than good. In August, they rejected the course on those grounds by a vote of four to three, with two abstentions.

This raised the ire of the new Stanford president Richard Lyman, who had been brought in to replace Pitzer. (The board had fired Pitzer largely for his inability to control the student unrest.) Lyman suggested that the Committee on Undergraduate Studies investigate the SWOPSI decision. He found it 'deeply troubling.' The committee mentioned the threat of violence, and Lyman rejected that out of hand. Shockley, he said, was no less qualified than a number of other SWOPSI instructors.[90] The American Civil Liberties Union jumped to Shockley's defense, in this case at his direct request.[91]

Despite the pressure, Shockley never did teach the course.

Other universities were not as forceful as Stanford in defending Shockley's rights. The Harvard Law Forum cancelled a debate between Shockley and Roy Innis, executive director of the Congress of Racial Equality, citing the possibility of disruptions and 'expressions of displeasure within segments of the Harvard community.'[92] Even Innis was upset, calling it a defeat for academic freedom. Innis blamed two black professors, Derrick Bell of the law school and Orlando Patterson, a sociologist, for 'playing a leading role in the suppression of academic freedom.'[93] Innis said he disagreed with Shockley 100%, but Shockley had a right to speak and 'the way to combat him is through ideas, not censorship.'

Yale was worse. Shockley was to debate William Rusher, the editor of *National Review*, but when the debate was to begin, 150 demonstrators, mostly minority students, jeered and booed for more than an hour until Shockley finally left the stage. The response to the incident by the Yale administration infuriated one of its most prominent alumni, William F. Buckley. Yale president Kingman Brewster had recommended that the undergraduate organization, the conservative Young Americans for Freedom, not invite Shockley. Brewster said it was not because of the threat of demonstrations but because he found Shockley's opinion obnoxious, not a reason that Buckley admired. Buckley was upset because he felt that unlike a Shockley debate with a liberal, which Shockley often won, he would this time have to debate a conservative. Rusher would not argue the specifics of genetics but the irrelevancy of the issue, an interesting point. But it wouldn't happen at Yale, apparently, Buckley lamented.[94]

Nevertheless, 12 Yale students were suspended. 'It makes me sick that even a small minority of Yale students would choose storm trooper tactics in preference to freedom of speech,' Brewster told the Yale alumni magazine later.[95]

Buckley thoroughly enjoyed the tumult, especially since it gave him another opportunity to hurl a rhetorical bomb at Yale's liberal administration. He sent off another column. He said Shockley was not harmless: 'he is a live carrier of scientific hubris' and encourages the Archie Bunkers in the world. On the other hand, how would the world look if everyone had the IQ of a Yale student – or of a Yale president?[96]

May died at a nursing home on 7 March 1977, at the age of 98. She had reasonable mental clarity and health for most of her later years and simply wore out. She had watched her son become one of the most important scientists in the 20th century and had seen him receive the Nobel Prize, and then she saw him become the Ishmael of the scientific establishment. Her loyalty never wavered, nor did her pride flicker. She had vowed her son would set the world 'right on something,' and as far as she was concerned, he did.

'I was very fond of May in many ways,' Emmy said. 'It was difficult to get an emotional reaction with her. She would say "yes" or "no." She was very analytically oriented.'[97] She also was sometimes brutally honest and insensitive. Like her son.

When May died, Shockley and Emmy kept her apartment for several months, trying to sort out her effects and deciding what to keep. Like her son, May threw nothing out. He learned much about his mother that he didn't know.

While there is some scientific data that paranoid behaviors are inherited, the Shockleys, mother and son, provide ample anecdotal evidence.

May had always been that way. The peregrinations with William in London, visiting scores of apartments before choosing, moving from

apartment to apartment, sometimes staying only a few weeks, their bat-
tles with their servants and landlords, their moves from house to house
in California, the distrust of the local schools; all are descriptive of paranoia: a moving target always is harder to hit. Most of all, however, was
her inability to sort through what is important and reject the rest. The
overwhelming quantity of things we all acquire in life wind up in trash
cans, landfills and furnaces. Not so the Shockleys. Everything had to be
saved. You might need it someday.

Emmy and Shockley found boxes and piles of papers on bookcases, in
closets, and in a desk with drawers crammed with stuff. May too saved
virtually every piece of correspondence she ever received. They found a
list of 41 names, friends who stayed over night in her homes, and a list of
those who had 'meals only.' They found all of her driving tests, many of
her surveying charts from Tonopah, and her son's Terman IQ test. In one
drawer, Shockley found a small pile of gold nuggets and jewelry from
William's collection. In other boxes, they found one of Alison's infant
gloves and the splinter that destroyed Shockley's dimple as a child.
Shockley discarded two large cardboard boxes weighing 90 pounds each.
He admitted it wasn't easy. Considering what he kept and gave to the
Stanford archives, one can only wonder what he threw out.

In remembrance, her son put together a 32-page fully annotated
booklet, a tribute to his mother's indomitable spirit written in his usual
turgid style. Inside he printed some of her poetry for the first time. It was
an act of love, but the word never appears in the text.[98]

None of her grandchildren came for the memorial service.

Jean died in 1977 after surviving a second cancer. She was diagnosed
with a third, separate malignancy, breast cancer, and it metastasized to
her brain.

She missed the final insult.

CHAPTER 13
'The high cost of thinking the unthinkable'

In the fall of 1975, Shockley returned to Gustavus Adolphus College for another Nobel conference.

He came with a clear agenda. He thought this was a great chance to force the issue on his peers. A few weeks before the session, Shockley wrote a letter to Polykarp Kusch in Texas, explaining what he planned to do. The letter was typical of Shockley's correspondence at this stage of his life: four single-spaced, typed pages, with multiple, numbered enclosures and end notes, all of it copied to the two dozen participants in the conference. Many other letters were even longer, almost to the point of the irrational.

'My opportunities to present reasoning related to dysgenics before competent scientific audiences have been few and far between since October of 1973,' he wrote. He did not intend to let the upcoming Nobel conference pass without taking the opportunity to make his point. Shockley had a new proposal, that each Nobel Laureate at the conference take a lie detector test. The question to be tested was: When you say that there is no racial difference in IQ, do you really believe it? Shockley had good reason to think some participants might 'flunk' the test. He assured Kusch he was not joking.[36]

Shockley had become a true believer in polygraphs, going back to his experience with the tack in the door at Shockley Semiconductor. He also had recently taken a polygraph test himself when several columnists challenged his sincerity. He 'passed.' This was just after Watergate, he pointed out to Kusch, and 'tests of integrity' were seen as desirable. That polygraphs are junk science never registered.

Shockley was in part reacting to the considerable number of scientists who supported his general thesis but would not say so in public. He had spoken to them on the phone; he had letters from them; he was told so in person, in quiet asides. Shockley knew he was not alone and that the support came from some highly respected authorities. Those supporters apparently felt discretion the better part of valor, no doubt encouraged to think so by what happened to anyone who dared agree in public. Even scientists who did not agree, but who did research Shockley could quote, or who defended his right to speak, or even appeared on the same stage in a debate, ran the risk of threats and figurative banishment from ivy towers. Author Berkeley Rice called it 'the high cost of thinking the unthinkable.'[101] Jerry Hirsch, his loudest opponent, at Illinois even went after Roy Innis in a character attack because Innis debated Shockley several times, and worse, seemed to enjoy it. Hirsch announced that Innis was an admirer of the notorious dictator Idi Amin.[81] He didn't say what relevance that might have had, even if true.

When the reporter Glenn Bunting challenged Shockley to prove he was not alone, he arranged for the reporter to talk to five scientists on the telephone with the promise that Bunting would keep their identities secret. They confirmed what Shockley told him. 'These supporters are not all retired academicians or conservative elitists with racial motivations, as portrayed by Shockley's critics,' Bunting wrote in *California Today*, the Sunday magazine of the *San Jose Mercury News*.

One, identified as a Harvard 'instructor,' said he agreed about the inheritability of intelligence, but said once you got into the matter of race, the issue simply became more trouble than it was worth. Another, based apparently in New York and described as a 'prominent medical scientist,' said essentially the same thing. There is a racial difference, he said, and 'you would have to be incompetent not to draw that conclusion. But you would also have to be out of your mind to draw such a conclusion in public. To take such an unpopular position is something most scientists are unwilling to do because they would risk having their government funding cut off and even jeopardize their jobs.'[102] *

Moreover, Shockley was forever having painful telephone conversations, such as this one, which he of course, recorded.

---

* Shockley thought it a matter of honor to keep the names secret when asked. When Bunting handed the notes for some of the telephone conversations to a secretary at the magazine for typing, he left the name of the scientist on the memo. Shockley, furious, ended the interviews and assistance with Bunting's article. Bunting wrote about it as an example of his irrational behavior.

In 1968, two scientists, known in this narrative as K and L (Shockley even kept identities secret in his files), both well known in their field and members of the National Academy, were giving a research paper at the academy that Shockley wished to quote. L had taken his name off the paper, although it was still listed in a meeting program, and K was to deliver the paper alone. Shockley called L because he wanted to quote from the paper and, he said, he wanted to make sure he had the attribution correct. He found that L was frightened to leave his name on the paper.

'Are you a coauthor of the paper? You appear on the program as a coauthor,' Shockley asked.

'I don't know how much importance that is. My name was taken off of there for reasons – not to get into difficulties,' L said.

'Difficulties with the university?' Shockley asked. 'But you are on the program and when K gives the paper, is he going to say "this is my paper?"'

'I suppose. Is that all right? He got all this work together basically,' L said, 'and this doesn't represent any original work and it didn't make much sense really to have two authors...'

Shockley interrupted. 'After talking to K, I assessed it up this way – I interpret that [as] suppression.'

'Well, this is a potential problem,' L said. 'I don't know how to get around it except it can cause potentially a lot of difficulties.'

'There are a lot of difficulties in the world,' Shockley said, with some exasperation, 'and I think they are going to be dealt with by getting it out in the open.'

When Shockley hung up the phone, he turned to his secretary and said, 'He's over a barrel.'

L wasn't the only one. Jensen also discovered underground and very frightened support.

Could Shockley have really thought the Laureates would agree to take a lie detector test about their racial attitudes? Was it just a publicity stunt? At this stage in his obsession, it is impossible to know.

The Gustavus conference was titled 'The Future of Science.' Shockley tried to convince delegates that a study of the genetic fate of humanity ought to be in science's future, that it would be the most important topic the Nobel conference could take up, and that Alfred Nobel would have approved. They clearly did not agree. Every time Shockley tried to bring up the subject, he was ignored as if he had not spoken at all. If he asked a question, it went unanswered. The more he tried to sidetrack conversations – he would bring up eugenics in the middle of discussions on solar energy – the more irritated they grew, finally groaning aloud and glaring at him. He had been silently voted out of the conversation and the community of his peers. It must have been painful to watch.

Anyone but Bill Shockley would have been at least disconcerted at this public humiliation by his confrères, but he showed no sign of it. Shockley by now perhaps was beyond embarrassment.

And the suggestion they all take polygraph tests? Unsurprisingly, none of the 30 Nobelists volunteered. 'An amusing exercise,' said fellow Laureate Julian Schwinger, 'but it's irrelevant. It hardly matters how sincere you are if you are wrong.'[100,103]

In 1971, critics of the 'hereditarians' had a new target, Richard Herrnstein, holder of the oldest chair in psychology at the oldest university in America, Harvard. Beginning with an article in *Atlantic* and a follow-up book a few years later, Herrnstein agreed that IQ was largely hereditary. He was unconvinced by the evidence of a racial difference, but was sure it was a factor in social class, which certainly had racial aspects. He believed that America had an IQ 'meritocracy,' a cognitive elite running the country, and he had some doubts this was a good thing.

Herrnstein, as one essayist put it, 'questioned the traditional liberal idea that stupidity results from the inheritance of poverty, contending instead that poverty results from the inheritance of stupidity.'[104]

He was added to the list of notorious miscreants. Although he barely mentioned race, he found himself being chased down the hallowed halls of Harvard by students calling him 'Nazi' and 'racist.'

Herrnstein was not a marginal character out at some flaky California institution with palm trees and coeds on roller skates in the middle of February. He was tenured professor in the heart of the Cambridge establishment. He found his most vicious critics up the hallway, down the stairway, or across the street. He blamed the reaction on a 'political orthodoxy on human equi-potentiality to which scholarship has become hostage' – essentially what Shockley was saying.[101]

Matters did not get calmer when Arthur Jensen reported research that he said showed that IQ tests were not culturally biased, one of the most persistent charges leveled against the tests. Many of his colleagues agreed.* Jensen said IQ was 70–80% genetic; Shockley said he could prove it was 80%; Herrnstein thought it could be a little less.

Some of the subsequent vehement opposition came from geneticists, both those who studied entire populations, and the microscope-wielding kind. Cavalli-Sforza, for instance, wrote in *Scientific American* in 1974, that the latest search into the human genome showed conclusively that genetic differences between individuals were much greater than the differences between groups of individuals, which would diminish the argument that there was a major difference between races. 'Race as a factor thus adds remarkably little to the differences we can detect between any two individuals,' he wrote.[105] Therefore, just studying the results of tests, as he assured Shockley in a debate, doesn't tell you about genetics.

This represents the rational – and vastly more interesting – part of the argument: the issue of perspective. In other words, Cavalli-Sforza and many of the geneticists were saying – to borrow a cliché – you can stand outside a forest, even walk around it with a clipboard collecting data on everything you can see, but you will learn almost nothing about individual trees that way. Quite true, except that neither Shockley nor Jensen ever said you could. Shockley always made it clear he was talking about *groups* of people, not individuals, and repeated, almost by rote, that there were many African-Americans who were intellectually superior to many whites. He encouraged them to go out and reproduce.

Conversely, Shockley, Jensen and Herrnstein would argue that you can walk through the forest with a clipboard, study everything about the individual trees down to the molecular level and learn almost nothing about the shape and texture of the forest. Also quite true, but the geneticists would counter that if you collect enough data you could rule out certain postulates about the forest.

---

* More than 20 years later, when Herrnstein, in collaboration with conservative economist Charles Murray, wrote the best-selling book *The Bell Curve*, the uproar was exactly the same, with the same cast of characters and the same result.

The fact that both sides are partly correct is the reason neither side can conclusively dispose of the other.

But most of the opposition lacked Cavalli-Sforza's good manners. They were less bothered by the science than by the perceived political, moral and social ramifications of the assertions, though they often couched their objections in scientific terms.

Part of what triggered their fury is the apparent logic of the hereditarians. You can't tell a dog breeder that intelligence is not genetic; they make their livings on that proposition.* There's a reason why retrievers and German shepherds, and if trained well enough, even stubbornly independent Siberian huskies, can serve as guide dogs. They are bred for memory, temperament and judgment. Conversely, Afghan hounds and Irish setters make lovely throw rugs because they are bred for beauty and stupidity. The notion of discussing breeding animals in the same breath as breeding humans, however, smacks of Hitler's Aryan delusions, and that quite properly upsets many.

Most people believe that smart human parents have smart children. Most of the people locked in this fierce debate were very bright people who probably had – or at least thought they had – very bright children, and you would probably have to search a bit to find one that didn't take at least partial credit. They just didn't want to do that in public.

On the other hand, laboratory tests have demonstrated just how close we are to our mammalian cousins. The genetic difference between chimpanzees and *Homo sapiens* is less than 5%, and that includes the opposable thumb. We revel in that 5%, and no one reveled more than Shockley. So, he asked, why shouldn't humans be subject to the same biological and genetic forces of other mammals in the other 95%?

We like to think ourselves above all that.

The issue goes deeper. If intelligence (or IQ if you will) is largely inherited, then all men and women clearly are not equal. If you believe that intelligence is a good thing to have, and that people highly endowed with such an attribute are better equipped to succeed in life, then are some people not inherently better off than others? You've now

---

* Every pedigree dog is descended from one pair of dogs, often less than 100 years ago, and has a very limited gene pool.

made a serious value judgment that goes against every egalitarian principle. We like to think we don't think that way.

Some of the opposition to Shockley and Jensen surely came from people who remembered the history of eugenics, and for them that was enough. Every time there is a justification for the notion that one group of people is superior to another, bad things have happened, including the Holocaust. 'Any investigations into the genetic control of human behaviors is bound to produce a pseudo-science that will inevitably be misused,' evolutionary theorist Richard C. Lewontin once said.[106] Indeed, if scientists found a gene for intelligence, he said, he didn't want to know about it.

Shockley often spoke of his optimism that people are good. If you lay out the facts, and encouraged free and open discussion, they will behave in a moral and decent way. His optimism, critics said, was probably misplaced.

If IQ is largely inherited, are we living lives far more deterministic than we like to believe? Do our genes doom us to a certain life, to certain behavior? Are genes destiny?

In the early 1970s, Edmund Wilson at Harvard helped to lead a new look at life called sociobiology. The theory held that social behavior in animals – and humans – stems from Darwinian natural selection. It is at least partly genetic. Attributes such as altruism, courtship and faithfulness extend across species, and Wilson and his colleagues began trying to find evolutionary, genetic explanations. Mild as all that sounds now, Wilson's theories were met with a ferocity that matched the antipathy toward Herrnstein, Jensen and Shockley. Essentially, it came from the same scientists.[107] It was not the basic idea that worried them, but the deterministic ramifications and how they might be misused.

The assault against Wilson was a sad time for American science. The attackers almost succeeded in driving him from Harvard and from scholarship.

Hirsch implied Wilson was anti-Semitic because he endorsed a sociobiology textbook that contained an anti-Semitic statement.[81] When a motion to censure sociobiology was only narrowly defeated at a meeting of the American Anthropological Association, Margaret Mead thundered 'Book burning, we're talking about book burning!' Fifteen scholars in the Cambridge area, including Steven J. Gould and Lewontin, formed the Sociobiology Study Group. They denounced sociobiology, linking it to Nazi ideology and racism. When Wilson spoke at a meeting of the American Association for the Advancement of Science (AAAS), he was physically assaulted and a pitcher of ice water dumped on his head.[108]

Sociobiology is now common currency, and Gould, at least before he died, seemed somewhat embarrassed by the treatment of his colleague. But he was just as vociferous against Shockley, Jensen, and Herrnstein, and later Herrnstein and Charles Murray. His main weapon then was his best-selling book, *The Mismeasure of Man*, and his column in *Natural History* magazine. 'Biological determinism is not simply an amusing matter for clever cocktail party comments about the human animal. It is a general notion with important philosophical and major political consequences,' he warned. The book was more widely respected by the public than it was by knowledgeable colleagues, who often noted that Gould managed to ignore 25 years of research that contradicted his arguments.

Another bitter opponent was the Princeton psychologist, Leon Kamin, who one critic said almost made a career out of attacking Jensen and Shockley. Martin Andrews, a psychologist at St John's University, described Gould and Kamin's attitude in the debate to that of theologians trying to purge heresy from the church of science.[110]

Guilt by association was elevated almost to an art form. All the opponents found the Nazi past of eugenics irresistible. All, especially Gould, were fascinated by the 19th century quacks who did everything from measuring head size to trying to determine criminal types from the shape of skulls, gleefully associating Shockley, Jensen and Herrnstein with those long-discredited souls.

Some of the opponents made statements in published papers that were arguably actionable. None went as far as Hirsch, who claimed that Shockley recruited Jensen to the cause, that Shockley was part of a conspiracy to foster segregation in the National Academy and in the pages of *Science*, and that he forced the NAS to capitulate – none of which is true. He called Shockley a 'dedicated, articulate and vociferous Nobel Laureate crusading for the cause of a well-financed network of powerful segregationists.' That was not true either. He charged the editors of *Science* with being part of the conspiracy by publishing papers fostering the heritability of IQ and suppressing papers that contradicted the theory.[81] None of that was true.

Even after Shockley's death, Hirsch would give him no peace. He wrote in a letter to *Science* that Bardeen told him that Shockley had a nervous breakdown in 1951, and was hospitalized for psychiatric care. Hirsch mentioned in the letter that Shockley was interested in psychiatry at that time and just happened to let fall the fact that Emmy was a psychiatric nurse.[111] Shockley did not meet Emmy until 1954; she and all three of Shockley's children deny that he ever had a nervous breakdown, and absolutely nothing in the voluminous, pathologically detailed archive lends any support to Hirsch's story.

The nurture supporters painted a picture, at least in the press, that virtually no one in science who knew about the subject supported research into race and intelligence, and that all thought such investigations ought to be either discouraged or outright banned. Not true.

In 1976, sociologists Everett Caril Ladd Jr and Seymour Martin Lipset published a survey in the *Chronicle of Higher Education* showing that 89% of American scientists opposed restrictions on research into race and intelligence, 62% without any reservations. Fifty-two per cent didn't think it even ought to be discouraged. Only 4% supported restrictions. Support for unlimited research was across the political spectrum: 28% of the most liberal and 26% of the most conservative scientists believed there should be no restrictions. Even 30% of engineers and agricultural scientists, who had nothing at stake, opposed restrictions.[104]

The idea that the hereditarians represented only a crackpot minority wasn't true either. In 1988, psychologist Mark Snyderman and political scientist Stanley Rothman published the results of a survey of 1,020 scholars knowledgeable about IQ. One of the questions in the survey was: 'Which of the following best characterizes your opinion of the heritability of the black–white differences in IQ?'

- Completely environmental – 15%.
- Completely genetic – 1%.
- Both genetic and environmental – 45%.
- Data are insufficient to support any reasonable opinion – 24%.
- No response – 14%.

In other words, a plurality of experts – almost half – believed in an IQ difference between races that was a combination of genetics and environment. Notice that the question assumes such a difference.

On the other hand, Snyderman and Rothman reported, one-third of all journalists and almost half the newspaper editors surveyed believed in a racial difference in IQ that was entirely environmental, and only about a quarter thought there was a combination of both genetic and environmental factors. Their coverage reflected that.[64] *Time* magazine, for one, reported to readers that Shockley's 'theories have been pronounced scientific malarkey by most experts in the field.'[112] (That wasn't true then; it isn't true now.)

Indeed, it was Gould, Kamin, Hirsch and Lewontin who represented the minority in science, although, as Snyderman pointed out in his book, you wouldn't know that from reading news magazines or newspapers.*

When recognized experts such as Jensen and Herrnstein took much of the attack, Shockley found himself growing less newsworthy. The argument had moved beyond him, and although he had become somewhat more extreme, he was essentially peddling the same old story. So he taught himself how the media really worked and became what is known in newsrooms as an 'operator.' He was a master at it.[113]

The first question an editor asks when deciding whether to assign a story is: is this news? If it is William Shockley giving the same speech on IQ and race that the newspaper has reported before, the decision is often no. Ideology or personal belief has nothing to do with it.

Reporters might have been somewhat relieved at that decision, however. For one thing, Shockley's speeches were considerably harder to cover than the average story, unless, of course, there were demonstrations, in which case reporters covered the demonstration, not the speech – presuming he was allowed to give the speech – and if he wasn't, well that's yet another story. Eventually, Shockley learned to turn the demonstrations to his advantage, and those sincere young men and women protesting his philosophy quickly became his best allies.

Without demonstrations, reporters had other problems. His speeches were crammed with numbers, charts, statistics, and the

---

* The survey was made in the 1980s. Snyderman and Rothman concluded: 'either expert opinion has changed dramatically since 1969, or the psychological and educational communities are not making their opinions known to the general public.'

jargon of at least two specialties. Few people know very much about genetics and fewer about statistics. The average general assignment reporter had no idea what Shockley was talking about and had serious difficulties providing context. If they tried – and they were obliged to try – to find opponents to Shockley's theories to balance the story, they frequently found those opponents equally obtuse, sometimes inarticulate and often emotional.

Shockley was a terrible speaker. Careful not to say something unintended, he carefully scripted all his speeches and then read them in a monotone, without inflection. Shockley also ignored the difference between written language and speech. His talks were written in exactly the style he would use for a scientific paper, no matter what the audience. He was never a good writer.

He may have established a modern record for flogging metaphors with one paragraph:

It is my intention to use significant members of the American press as the blocks or pulleys... and the First Amendment as a line upon which I shall endeavor to exert a force so as to deflect the rudder of public opinion and turn the ship of civilization away from the dysgenic storm I fear is rising over the horizon of the future.

This kind of language is hard enough to understand on paper. It is impossible when read aloud quickly. You can always stop, go back and reread a written paragraph; you can't do that sitting in an audience listening to an oral presentation. Shockley was so bad that Larry L. King wrote in 1973: 'Dr. Shockley... made such an inept presentation that he probably could not have instructed us how to catch a bus.'[46]

Debates, where the potential for verbal fireworks seemed attractive, quickly disappointed. Shockley stuck to the script no matter what anyone else on the stage said, and his opponents usually and quickly resorted to cheap emotional diatribes that left them looking not much better. Debates did contain useful quotes, or sound bites, but lacked the skeleton upon which to hang them. Writing a story with all quotes and little substance is not easy, even for television reporters.

More than most professions, journalism has an active gossip system that even runs to competitors and across geographical areas. Word got out quickly that covering Shockley, unless he was disrupted, was very hard and no fun at all, making reporters reluctant to volunteer. Most reporters love stirring up folks – it's one of the reasons they became reporters in the first place – and most editors were once reporters. But the amount of aggravation they took for a story with questionable newsworthiness exceeded the benefits. So, they began finding reasons not to run Shockley stories. Shockley learned how to finesse their indifference.

Manipulating the media is quite easy once you understand its internal functioning and separate yourself from all the ideological myths. Holding colorful or, especially, violent demonstrations manipulates the media, as every civil rights, anti-Vietnam war, and environmental protester quickly learned. This is especially true for television, where pictures beat words every time and not a lot of thought is given to content. You learn when the deadlines are and time your demonstration so that it is far enough away from the deadline to get covered but close enough to minimize the amount of thought behind that coverage. That's why all demonstrations begin between two and three in the afternoon, to guarantee coverage on the five o'clock news and make the first edition of newspapers.

Shockley went far beyond those simplistic tricks. He became so good at manipulating the media that Rae Goodell, in her classic study *Visible Scientists*, used him as an examplar of the operator. He was helped by sympathetic editors across the country.

Shockley learned that the Associated Press, the largest news service in the world, was – and is – a cooperative, owned by its member newspapers and broadcast outlets. If a member asks that a story be covered, the AP usually complies. No AP bureau chief wanted to have to answer to the New York headquarters if a member complained that coverage was declined unless there was a very good reason.

Say Shockley had a speech scheduled in Cleveland and wanted to make sure it went out on the AP wire. He would call a friendly editor in, say, Alabama, and tell him about the speech. The editor, probably because he really wanted to print a story about the speech, but possibly just to do Shockley a favor, would call the local AP bureau in Montgomery or Birmingham and request coverage. That AP bureau would send a Teletype message to Cleveland repeating the request, and if he could, the bureau chief in Cleveland would assign a reporter. This was normal business; AP was founded to provide such a service.

The AP reporter would come back to the bureau with the story. It could be sent privately on a backchannel line to the newspaper in Alabama, but every wire service reporter and bureau chief lives under the axiom that nature abhors a quiet news wire. It would be just as easy to send the story on a regional wire or even the A wire, the main national and international circuit, so everyone (including their New York bosses) could see it.

This hardly assured Shockley that anyone would use the story, but it did assure him that wire editors (then called 'cable editors') on the national desks of other newspapers would see it. Who knows, maybe that editor was sympathetic too, or loved to stir up readers, or it was a quiet news day, or it was before a holiday and they had a huge 'news hole' to fill, or another assigned story didn't come in on time, or.... Shockley played that card unashamedly. It worked most of the time, and helped keep him in the news.

Plus, Shockley had an ally at Stanford. While it was an unlikely marriage, it was, from Shockley's perspective, made in heaven.

Bob Beyers came to the Stanford University News Service from the University of Michigan in 1961, brought in by Lyle Nelson, who developed the notion of public relations offices for universities. Nelson – about as close to being the patron saint of institutional public relations as anyone is likely to get – developed the unique philosophy that the best public relations strategy is to tell the truth, all the time – all of it, even when the truth is painful. While embarrassment may be excruciating in the short term, it pays off in the end. It is Nelson's model that is responsible for the often first-class public relations of American universities and colleges.

Beyers believed in Nelson's model passionately, and he hired newspaper reporters instead of public relations people as writers in the news service with instructions to cover Stanford just as they worked for a local newspaper. He gave them complete independence. He and Nelson even encouraged investigative pieces, essentially biting the hand that fed them, and, with Nelson's considerable clout, ardently defended the stories when they were challenged. Beyers refused to let university officials read stories before they came out, so there was no pressure to censor them. He kept arm's distance from the development office, which usually runs public relations offices at universities, because he thought any link between the news service and the office that raised money for the school was a conflict of interest. Beyers was probably the inventor of the preemptive press release. If anything bad were about to break, he would rush out a press release before the newspapers had the story. 'By getting the story out first, you actually minimize the coverage,' he said, and he could demonstrate countless instances when that's exactly what happened.

The result was that the Stanford University News Service had a unique reputation among university PR operations and still is held up as the model for how these offices should ideally work. Few if any others succeeded – or were allowed to try – and even Stanford no longer operates that way. But it did then.*

Normally, a scientist like Shockley was under the watchful eye of the news service's science writer, then the veteran Bob Lamar. But Lamar grew to detest Shockley and would have nothing to do with him.**

Shockley and Beyers were an unlikely pair because Beyers probably was greatly offended by much of what Shockley said. In the early 1960s, Beyers volunteered to do public relations for the 'Mississippi Summer' civil rights movement, a job with some physical danger. He did everything he could to bring diversity into his office. But he and Shockley were alike in two important respects: both passionately believed that you put ideas out to the people and rely on the public to make the right decisions, and neither man could draw any proper distinction between work and life. Being called by a furious, frightened or inquisitive faculty member on a weekend or in the middle of the night was just fine with him.

Often asked how he could continue to publicize a man whose words probably nauseated him, Beyers said: 'The man deserves to be heard, and if you give a voice to him, there's a responsibility to get it out fast and get it out accurately so that the controversy has a better chance of being based on what he actually said, than to have some misrepresentations of what happened.' He and Shockley never discussed racial

---

* While Beyers' honesty played well with reporters, who trusted the news service implicitly, it naturally earned him the enmity of Stanford administrators and bureaucrats, people who found themselves clearing their desks when Beyers entered their office to prevent him from snooping and printing what he found. His sometimes painful honesty eventually cost him his job.

** I was hired by Beyers from the *Philadelphia Inquirer* to replace Lamar when he retired in 1979. Since Shockley and Beyers had a working rapport, Beyers continued to handle him and I had very little to do with Shockley.

theories. Beyers simply told Shockley that he was going to accurately report what Shockley was saying 'and it's not going to be me injecting my views.' Beyers made sure that the other side, particularly when Shockley was debating African-Americans, got equal play in the press releases.

He and Shockley got along famously, and Shockley now had America's best institutional public relations office at his disposal, complete with mailing lists, archives, lists of media sources, and a reputation for integrity. This probably kept him in the public eye longer than he would have on his own.

Shockley had a strategy for getting media attention – 'and it was very clear to me that it was a strategy,' Beyers said. For instance, he knew that he got more coverage when he debated African-Americans than when he shared the platform with whites, so he tried to entice them into debates. He learned quickly how to milk demonstrations into maximum coverage.

'He was obsessed, obsessed with these ideas,' Beyers said of Shockley. 'He was particularly obsessed with the attitude that the National Academy had taken, and speaking personally, I thought that was a pretty stupid thing that they did.' Shockley was determined the National Academy was not going to gag him. Beyers, who was committed to unfettered science and open communications as articles of faith, could enthusiastically help. 'That goes to the soul of what I believe in terms of the job that I did here, which is that no matter how repugnant something is, you get it out there and let other people get a shot at it. Because they will.'

Shockley was known as 'Dr Beep Beep' in the news office in honor of his telephone recording device, and Beyers always warned reporters asking to speak to Shockley they would be taped. No one ever objected. Although Shockley was listed in the phone book, most of the reporters calling him went through Beyers.

Shockley often used Beyers as a sounding board and sought his advice. Giving faculty members advice was part of Beyers' job, and many took advantage, including Shockley's opponents, such as the biologist Paul Ehrlich. When a speech in Kansas was disrupted, Shockley called Beyers to make sure Stanford sent out a release. One day, before he learned what a boon demonstrations were for publicity, he called Beyers to ask him if he should cancel a talk in San Francisco. There were going to be demonstrations. Rather than give advice, Beyers laid out the alternatives and what would happen in each case. The speech eventually was cancelled, so the conversation became moot, but Shockley learned to rely on Beyers' wisdom. And he never walked away from a demonstration.

When the speeches became redundant there was always the occasional stunt to keep interest. So it was that in 1980, Shockley called Beyers and told him that the *Los Angeles Times* was going to run a story that he had masturbated for the sake of his principles; he had contributed to a sperm bank. Beyers decided it was time for a preemptive press release.

CHAPTER 14
'I love you'

Robert Graham made a fortune as the inventor of shatterproof eyeglass lenses, worn by millions of people around the world. He was a eugenicist. In the early 1960s, he decided to put his money where his beliefs lay, and he founded a non-profit sperm bank, along with gloomy Herman Muller. What he eventually called the Repository for Germinal Choice was located in a converted pump house on Graham's 10-acre ranch in northern San Diego County until it was moved to the basement of a nearby house.

At first, he solicited only Nobel Laureates, writing them letters, essentially inviting them to jack off into metal tubes that would be inserted into frozen nitrogen. The plan was to offer their sperm to worthy mothers, which by his definition was any intelligent woman who could raise a child in a healthy home.

Since sperm, frozen solid, can last 1,000 years, it would be possible to sire a child who would live into the 30th century. To Graham's surprise, only three Nobelists found that enticing.

Most didn't answer the letter. Graham would later conclude that most Nobelists were too old for such things (and most women didn't want old sperm), or not much interested in such a plan, so he expanded the donor base to healthy men of achievement with IQs of 130 or higher. Most of the donors turned out to be scientists. None was African-American, although Graham would have been happy to get their sperm as well, he said.

The announcement of his sperm bank set off a wave of criticism, with charges that Graham was playing God, that his plan was elitist and racist, and that it was scientific nonsense. Intelligence in humans, unlike golden retrievers, can't be reliably reproduced, the critics pointed out. Graham agreed he could not guarantee a brilliant child. He thought he could lessen the chances of producing a dullard.

For a $500 deposit, Graham would ship, by Trailways bus, a tank holding a three months' supply of sperm from six donors, identified essentially as 'Purple,' 'Green,' 'Blue,' etc.[115] The mother would have a catalog with some description of who contributed the sperm. One woman, for instance, selected 28 Red, a man described as a handsome, blond, athletic science professor who scored 800 on his math SAT tests. She produced a little boy named Doron Blake, who is the only identified child of the program. By the time Graham died in early 1997, he had produced 218 children in five countries.[116] None are known to be geniuses. According to the bank all were bright and doing well.[117]*

No one knows if any of the Nobelists sired children.

The only one to identify himself was Shockley. He was contacted by Ed Chen, then the science writer at the *Los Angeles Times*, who asked permission to identify him. Shockley agreed – probably without too much reluctance, possibly even having instigated the disclosure – if Chen agreed that the story would contain one paragraph in which Shockley explained why he did it. Shockley and Chen would collaborate on the paragraph. Chen accepted the deal and the story contained the following.

I welcome this opportunity to be identified with this important cause. But I want to make it clear also that I don't regard myself as the perfect human being or the ideal candidate. I'm not proposing to make supermen. But I am endorsing Graham's concept of increasing people at the top of the population, which is to be differentiated from anti-dysgenics – my past and present emphasis on reducing the tragedy for the genetically disadvantaged at the bottom.[118]

Beyers sent out his release. To make sure the word got out, Shockley sent off one of his own, on the cluttered and slightly mad-looking stationery of a foundation he formed to support his research: the Foundation for Research and Education on Eugenics and Dysgenics (FREED). The tax-exempt foundation had numerous small supporters, but essentially it consisted of Shockley and Emmy, and mostly it was his money.

There is no reason to think Shockley contributed his sperm purely as a publicity stunt, but he saw it as a fine way to get his name in the newspaper again. It worked.

---

* The sperm bank closed in 1999.

The reaction ran from appalled to amused, not what he expected. San Francisco's great wag, Herb Caen, wrote 'Shockley's donating sperm is proof that masturbating makes you crazy.'[119] In a now-rare moment of humor, Shockley, appearing on one television show, took off his jacket, spun around and asked the audience if he looked like a superman. On another, he gleefully reported, a television audience, given the choice of his genes or Elvis Presley's, voted 65% for Shockley, 4% for Elvis.[120]

On 29 July 1981, an African-American science writer at the *Atlanta Constitution* named Roger Witherspoon, wrote a column headlined DESIGNER GENES BY SHOCKLEY. He compared Shockley's idea of a voluntary sterilization program to the Nazi experiments on the Jews.

He said Shockley told him the only objection he had to the Nazi experiments was that they were aimed at the Jews, 'the most intellectually advanced segment of their population,' Shockley was quoted as saying. 'It was anti-Jewish. In that, they [the Nazis] made a mistake, in my opinion.'[121]

The article was reprinted a week later in a Palo Alto newspaper.

Shockley, by now largely hardened to insult, thought that was profoundly offensive – and a serious misrepresentation of his beliefs. For the only time in his life, he sued for libel. He sent out a press release announcing he had filed a $1.25 million libel suit against Witherspoon and Cox Enterprises, owner of the *Constitution*, and had hired fiery Atlanta lawyer Murray Silver to represent him. 'The article contains the most unwarranted derogatory presentation of my position that I can remember,' Shockley said in the release.[122]

It took three years for the suit to come to trial in Atlanta. Journalists circle the wagons when one of their own is under fire in a libel suit, but many thought Witherspoon's column was over the top and kept at arm's distance.[123] Shockley's lawyers used all three of their challenges on prospective African-American jurors and had none left to block a black woman from the six-member jury.[124]

Witherspoon stood by his column, pointing out that it was an opinion column and that was his opinion. He did not have malice toward Shockley; he only disagreed with his views.[125]

The trial lasted eight days. On 15 September 1984, the jury agreed after three-and-a-half hours that Witherspoon had libeled Shockley and then awarded him one dollar in actual damages and nothing in punitive damages. Essentially, the jury agreed that Witherspoon's column met the standards of defamation, but that by then, Shockley's reputation wasn't worth very much. Shockley, left with huge legal bills, was stunned, even if his lawyer did call it a 'splendid victory.'

Before the trial, Witherspoon and Cox's attorneys suggested that Shockley sued to get publicity, and at least one reporter noted that Shockley's time in the media seemed to have run out. That was true.

Shockley had managed to offend almost every reporter asking for an interview. He wanted to control what was written about him, or at least to make sure he was dealing with reporters who understood what he was saying, but he took it to such extremes that many elected to pass on a story.

Shockley, using techniques he refined during the war as a weapon against the U-boats, said he could prove statistically that intelligence had to be 80% inherited. Anyone who asked for an interview had to sit through a demonstration of this proof using a deck of 50 specially marked cards. Shockley had reduced the inheritability of IQ to a card trick.[101]

Most reporters were lost and most of them didn't want to take a crash course in statistics for a story. Those following along so far discovered that was only the first step.

Freelancer Michael Rogers, doing a piece for *Esquire*, was first asked to spend several hours in Shockley's outer office performing the card trick with a grad student. Then Emmy, who gave him two pounds of papers to read from the prepared packets, interviewed him. At Emmy's recommendation, Shockley called Rogers and invited him to their house, where Rogers found himself being examined again. Shockley played taped telephone conversations with unsympathetic reporters and watched to see Rogers' reaction. Later, in a hostile 45-minute telephone call, he tested Rogers' reaction to a television tape he asked the writer to watch, and when he was dissatisfied with Rogers' answer, 'flunked him.'

He sometimes asked writers to do work for him as the price of an interview. When Rae Goodell refused, he threw her out of the house. 'You're not using me,' he called from the doorway. 'I'm using you.'

For most journalists, offended by this treatment, Shockley had become more trouble than he was worth. But not to all.

In 1974, a Minneapolis-based medical writer, Syl Jones, wanted to do a piece on the nature–nurture debate for *Modern Medicine*, a publication aimed at doctors. Jones had made a study of Shockley's work, so when he telephoned Emmy he was ahead of most of his colleagues. Shockley and Emmy, as they often did, analyzed the tape of the conversation over dinner before agreeing to continue contact with Jones. Shockley, however, insisted that Jones undergo a series of telephone tests on statistics, and that he fill in details of his background, down to his family and schooling. Shockley finally agreed to an interview, and in October 1974, Jones and a freelance photographer showed up at Shockley's home. When Shockley answered the door, he saw a black man and a white man and reached out his hand to shake the white man's hand, and said 'Hello, Mr Jones.'

Jones was the African-American.

Shockley seemed stunned at first, calling him 'the exception that proves the rule,' and insisted Jones take one more test, this one on how the Pythagorean theorem related to a long-forgotten part of Shockley's dysgenics theory. Jones said he somehow 'passed.'[126] He got the interview.

Although Jones contacted many opponents to the race-IQ theory, he essentially devoted the story to Jensen, with some Shockley and Herrnstein. Entitled 'Thinking the unthinkable about race and IQ; Are racial differences a real factor in intelligence?,' the article was unusual for two reasons: it was not the kind of article readers expected in *Modern Medicine* – which was more likely to print something about the latest uses of ultrasound – and it was unusually fair and well researched. Indeed, it may have been the best article written on the subject, complete with charts and graphs explaining Shockley's card trick. The piece was fair to a fault, critics said, and the magazine was almost submerged in responses. They ranged from outrage that *Modern Medicine*, of all publications, would print such a piece ('I am appalled that *Modern Medicine* should publish so biased a discussion of a highly controversial subject' – Philip Handler of the National Academy), to offense ('The article is an affront to the medical and scientific training of *Modern Medicine*'s readers' – Marcus Feldman of the Stanford genetics department), to approval ('The author of the article should be congratulated on providing a fair and informative discussion' – Everett R. Dempster, professor of genetics, emeritus, UC Berkeley). Same article. The letters published were about half for, half against.[101]

Shockley kept in touch with Jones, obviously respecting the reporter's ability to understand the statistical skeleton of his theory. When, in 1980, Shockley was considering publicizing his contribution to the sperm bank, he called Jones for advice.

So, that year, when Jones sold the idea of an interview to *Playboy*, Shockley was happy to comply. It would be the most famous, and except for the interview for *U.S. News & World Report*, the most controversial one he would ever give.

Shockley was confronted – as he rarely was – by someone who knew what Shockley was talking about. Jones had the tenacious instincts of a journalist and a keen sense for the jugular. Jones was wise enough to push Shockley and then let him hang himself. The interview, published in *Playboy*'s August 1980 edition, was Shockley's last best chance to get his points across, and he did. The interview also ended any lingering doubts about Shockley's attitude about race.

He called what he was proposing 'raceology,' the study of races, not racism. But he clearly was not coming at it from a purely unbiased angle, and Jones would have none of it. 'The smack of racism attributed to "my rhetoric" lies in the ears of the listeners,' Shockley said. 'It is not present in my written or spoken words. The word "racism" carries with it a connotation of belief in the superiority of one's own race, plus fear and hatred of other races, and lacks any hint of humanitarian concerns.'

'You believe quite simply that whites as a race are superior in intellect to blacks?' Jones said.

'Statistically, yes. But not in individual cases.' Shockley repeated his line that there were many blacks superior to whites, and that the white race wasn't necessarily the superior race.

'How do you feel about prejudice?'

'Prejudice that is not supported by strong facts is both illogical and not in accordance with truth. The general principle that truth is a good thing applies here.' On the other hand, he said, if it turns out there's sound statistics behind those feelings, well then prejudice might not be an evil – it's not, by definition, prejudice. If you found a breed of dog was unreliable and temperamental, why shouldn't you regard it in a less favorable light?

He quoted from studies of intelligence tests on blacks in Africa showing that they scored universally below average. None, he said, came within 10 points of his estimate for IQs of blacks in California, again bringing up his theory that the brighter the black the more white blood he or she carries.

He did not oppose racial intermarriage as such, but 'if you pick two black people at random in the black population and mate them and produce children, and you take two white people at random in the population and mate them and produce children, the existing statistics fit into the pattern that I call an inescapable opinion that the black children will be, as far as the IQ tests are concerned, inferior to the white children.'

Jones asked if he had any contacts with blacks in his life. Shockley admitted very few, an occasional maid here and there, and not all of them were very good at being maids. When his kids were in school in Madison, he pointed out, the president of the high school student body was an African-American. 'I thought that was a constructive social development.'

'In 1961, my wife and I were in a hospital for months in casts after a head-on collision. Most of the nurses who took care of us were black, and the quality of their care stood in marked contrast to that of the white nurses. My wife and I were most impressed.'

'What was it that impressed you so highly?'

'They gave us the best care and were the most natural and comforting that I had. In fact, while my cast prevented me from doing so, they were the ones who cleaned my rear end properly,' he answered.[126]

As stunning as the whole interview was, it was one remark, buried early on in the transcript, would become the most famous. Jones, proving an axiom in journalism that often the most innocent questions get the best quotes, asked Shockley how his children turned out. Shockley answered:

In terms of my own capacities, my children represent a very significant regression. My first wife – their mother – had not as high an academic achievement standing as I had. Two of my three children have graduated from college – my daughter from Radcliffe and my younger son [Dick] from Stanford. He graduated not with the highest order of academic distinction but in the second order as a physics major, and has obtained a PhD in physics. In some ways, I think the choice of physics may be unfortunate for him, because he has a name that [he] will probably be unlikely to live up to. The elder son is a college drop out.

Most of the world read that as a slap at Jean for producing inferior children. Shockley blamed that interpretation on a press release put out by *Playboy*'s PR department, and fired off a series of letters to the editors of newspapers that ran a story, including the *New York Times*, which had editorialized against Shockley. 'Regression,' he said, referred to regression toward the mean, the statistical phenomenon that explains how, within a generation, there is a tendency for the individuals in a population to be closer to the mean than the previous generation. He said not one reporter got it right, or apparently knew what regression to the mean was. 'For both the high and the low IQ parents, the children tend to regress towards the mean or average of the population,' he said. He pointed out that May had an IQ of over 150 and he, with an IQ of 130 or less, was a regression to the mean.

'Not one of about one hundred newspersons who dealt with this story was enlightened enough on facts about heredity and intelligence to correct this error [of misinterpretation], and other examples of the dark-ages dogmatism that dominates the views of the American intellectual community.'[128] It is probable he meant a regression to the mean, but if he had not added the sentence about Jean, more people might have believed him – including his children.

He would never live that comment down.

Not counting any unknown offspring from the sperm bank, Shockley had only one grandchild. Dick married a Japanese woman and they had a daughter, Hanna-ko. After a divorce, her mother took the girl back to Japan. She is now a college student in Australia, estranged from her father. Dick does not remember his father paying any particular attention or showing any affection to his one and only grandchild, but Shockley left money for her in a trust fund.

Neither Alison nor Bill had children. Dick and Alison saw their father occasionally, Bill never. Alison says she never realized just how dysfunctional her family was until she moved to Maryland, next to a fine, close, loving family, with whom she became friendly.[129]

Ironically – and thanks largely to Emmy – Shockley's home life was stable, and if, by romantic ideals, not perfect, it was happier and more secure than many. Shockley depended entirely on Emmy. He was loyal and faithful. He called her 'Mrs Shockley' to others. He was romantic in his own way, often sending flowers, and she apparently was a sucker for flowers. She adored him, and almost never regretting a moment. 'We had fun,' she says often, using a word not usually associated with Bill Shockley. To Emmy he was and is a great man. She happily devoted her life to him and would not listen to unbelievers.

When they met, Emmy was a very bright woman in a world that undervalued very bright women. She was headed for a productive but unglamorous career in Columbus. Less than two years later, she was dancing with the King of Sweden at the Nobel ball, and beginning a life traveling the world. She made her bargain, and she did not look back.

She and Shockley did have some bad moments, she confessed. The arrangement had been from the start that if either of them wanted out of the marriage, all they had to do was say so. Twice Emmy felt it was time to end it, she says, but she said later she could not remember why. She describes the incidents as 'disagreements.' During the first incident, as they sat in the family room, he said nothing, just shook his head. They later agreed they would try again, and they did. The second time, she remembers, was at the kitchen table. She never left, though. He apparently never brought up the subject. He was smart enough to know he needed her.

Shockley worked out of the last two bedrooms at the end of a long hallway. They slept in the front bedroom, just off the entrance hall. When he finally had to abandon his campus offices, he moved the mountain of material to the house, stacking up cardboard boxes in the garage, in two safes, and in both offices.

The house was always neat, orderly and immaculate – except Shockley's office, where stuff was piled all over: on the floor, on the desk, on shelves. Tape cassettes were everywhere, as were magazines, clippings and notebooks. His photocopy machine, worked almost to the point of metal fatigue, sat opposite his desk, ever handy. 'What can I do,' Emmy asked to one reporter? 'I don't tell him to clean it up. He's the one that bosses me around.'[102]

Most of his electronic equipment came from the local Radio Shack store. His favorite was a clock radio that projected the time on the ceiling over his bed. He regularly listened to a radio talk show from Chicago in the middle of the night, and he could keep track of the time by watching the ceiling. His 'security blanket' was his portable Sony tape recorder.[102]

Most of the time, when Shockley and Emmy were not working, they were either outside on the back deck or in the family room, which was lined with books. The centerpiece of the room was a large television set and a well-used videotape player. A good day ended over drinks outside, if they could; in the family room when they could not.

May's paintings and those of one of her friends were the only art in the living room. A small table next to the main door, looking a bit like a shrine, was a collection of awards; shelves across the entrance way held a display of objects, including a model of the first transistor.

Shockley was often in the garden and confessed a minor fascination with the physics of sprinkler heads, even writing a treatise on them. He drove around for almost ten years in a 1974 Fleetwood Cadillac, but they mostly used a small Chevrolet Chevette.

They had very few friends left. He had driven the rest away and frankly did not miss them, he said. 'The type of people I am drawn to are those who have similar views to my own,' he said. 'These views are the main focus of our activities. These are not ones in which any particular teamwork has developed with other people.'[102] He was not likely to make many new friends, largely because of his famous propensity to inviting guests to dinner and leaving them at the table to return to work in his office down the hall, letting Emmy pick up the pieces, as she did with Jensen.[82] He was not a gracious host.

Until his health began to fail, he made constant use of Stanford's two swimming pools, where he was famous for insisting that everyone in the pool was in a race with him whether they knew it or not. Others would jump or dive into the water, do a few laps, and get out. Shockley saw anyone in the pool as competition, swam along side them and started racing them, sometimes taunting them to challenge him.[130] Once, when Gibbons, half his age, wouldn't race, Shockley sneered, 'Chicken!' and swam away.[131] After a while, a few waited until he got out of the pool before getting in to avoid him.

Emmy never saw his insensitivity, and seemed incapable of interpreting how he behaved as ill-mannered or unbalanced. He was not, she insists, insensitive to her, quite the opposite. 'He would know that I was feeling bad about something and he would do something about it. He wouldn't talk about it,' she says. 'I'd talk first; he'd just do something. His sensitivity and perceptiveness are very important qualities of his,' she said, slipping into the present tense as she sometimes did almost a decade after his death. 'He was a very warm, sensitive, perceptive person.'[41] Her opinion of his sensitivity puts her in a minority of one, and his former secretary, Mary Clouthier, contradicts her description. Clouthier says that in the 1960, Emmy's hearing began to fail, and often Shockley would tell her to do something that she did not hear. Unaware of her hearing loss, apparently – or possibly just insensitive to it – he would yell at her on the telephone because she had not followed his orders. Clouthier says the painful scenes were a reason she eventually quit.

But that surely is not the image of Shockley that absorbs Emmy. To her, he was the great romantic and theirs was a true love story, even if their picture of love and the rest of the world's sometimes were out of register.

'I don't know what you mean, really, by love. We had our own definition about what it was to be in love. It was no "enchanted evening" thing ever. When it started, I really don't know. It was gradual, over a period of time. Enjoying him, talking with him, doing things with him, having a lot of fun. It was a good relationship and a good experience.'

Well, there were some enchanted evenings. The night before his 70th birthday, they made love grandly, she says, and the next morning, when she went to the office, she found a corsage of seven small, beautiful orchids. With it was a card, 'better than ever at seventy.'

Did he ever, in the more than 30 years they were married, tell her he loved her?

'I remember vividly,' she says, 'March 7th, 1984. We always went outdoors for dinner. I was headed back to the house with the salad plates and he called out to me. I turned around and looked, and he said, "I love you."

'I just stood there. I didn't say anything. I stood there and looked at him. I think he knew.'

―≺―

By the early 1980s, the world's attention drifted to other things and to other people. Scholars such as Jensen and Herrnstein had drawn most of the fire in the nature–nurture dispute, and Shockley was seen more as an extremist – at best, an eccentric.

He tried several stunts to keep in the media. The tricks were less driven by ego than they were by a sincere desire he had to keep the issue before the public. He considered himself at root a public relations man, he said often. If the value of a public relations man is judged by how much play he got in the media, he had been successful for 15 years. If the value is judged by how many minds he changed, he was an abject failure. He probably drove away far more people than he drew. Jensen called his personality 'reverse charisma.'[62]

In 1982, he ran in the Republican primary for the US Senate seat being vacated by S. I. Hayakawa. His one and only issue: dysgenics. He used his own money and showed up when invited, but he was largely and sometimes ostentatiously ignored. He didn't seriously think he would be elected, but it was a way of getting his issue before the public again, and getting his name in the newspapers. Shockley came in last. He got more publicity with his idea that presidential candidates take polygraph tests to see if they are telling the truth in their campaigns.

He had become a man who couldn't be invited for freshman orientation with Stanford's other Nobel Laureates without having the Stanford administration apologize to the freshmen.

His racism destroyed his credibility. Almost no one wanted to be associated with him, and many of those who were willing did him more harm than good. Partly that was his fault. He was incapable of drawing lines, of declaring that some people were worthy allies and others would discredit him just by association. Some of his allies were people that no moral, thinking soul would ever be associated with. His archives contain letters from white racist groups, including the White Citizens' Council of Mississippi, which tried to associate itself with his research. While he did not embrace them, he did not discourage them as he should have.

Shockley did seem to draw the line around monetary contributions. He turned down donations from the racist groups, but not necessarily from all racists. He carried this policy to its extreme in 1986, the last overseas trip he and Emmy took. He had been invited to South Africa to give a lecture on the transistor. This was at the time when most of the world was boycotting the apartheid regime. Shockley was happy to go, but he insisted – demanded as a condition – that he be allowed to give a lecture on dysgenics.

'He couldn't say no,' Emmy says. 'Don't ask me why.... They [the South Africans] didn't want him to do it.... He felt they should be doing some research on the subject as well.'

In the *Playboy* interview, in his letter to the Pioneer fund describing research that might reduce 'non-white' births, in his theory on white blood, in his much-expressed belief in the intellectual superiority of the white race, he certainly fit anyone's definition of racist.

Richard Goldsby, a black chemist, and one of the few African-American scientists to take Shockley seriously, defined him best. He called Shockley a racist, but not a bigot. 'He's a racist because he thinks he can make statistical prediction of behavior by population. He's not a bigot because he apparently does not despise blacks.' 'I like to debate Bill, because I always win,' Goldsby said.[102]

Shockley presented his sterilization plan as a 'thinking exercise,' but he always made it clear he knew how it would turn out. He always said he would be pleased if serious study showed he was wrong – 'if proven wrong my distress over a scientific setback would be more than compensated by the fact that the new findings would be of great benefit'[119] – but no one could doubt that he was sure he was right.

Ironically, as Shockley became more discredited, the science moved, some might say, in his direction. Although there are still a few holdouts, Wilson has been vindicated. Genetic factors are now believed to be largely responsible for behaviors such as traditionalism, stress reaction, absorption, alienation, feelings of well-being, harm avoidance, aggression, achievement, control, and social closeness.[132] So too are extroversion, conformity, some forms of creativity, worry, optimism, cautiousness, orderliness, intimacy, and yes, paranoia.[133] And intelligence.

No one believes Shockley's claim that IQ is 80% inherited, but figures range from 40 to 70%, with most scientists ceding the point that genes account for at least half, probably more. Improving the environment is a moral imperative, but there are limits to what environmental improvements can do, many now agree. While it still is dangerous to say so in public, the genetic component must be considered if lives are to be improved.

Much of the view of intelligence comes from both the laboratory table and the studies of identical twins separated at birth and raised in different homes (environments). Shockley cited the twin studies often.

The best known are those of Thomas Bouchard at the University of Minnesota. Unable to get much support for his research at first, Bouchard had to turn to the Pioneer Fund, although the National Science Foundation eventually chipped in. Partly emulating the data collection techniques of the eugenics lab at Cold Spring Harbor, the Minnesota team gathered every possible bit of information about sets of twins, all of whom had been raised separately.

Despite their obvious implications for the nature–nurture debate, the Bouchard studies are very popular with the media because of the positively spooky similarities found among the twins. For instance, the famous 'Jim Twins': Jim Springer and Jim Lewis were adopted by separate working-class families in Ohio as infants but never met. Both were found to like math, mechanical drawing and carpentry, and were bad spellers. Both worked part-time as sheriff's deputies; both vacationed in Florida, both drove Chevrolets; both had dogs named Toy; both married and divorced women named Linda and then married women named Betty; they named their sons James Allan and James Alan respectively; they had identical drinking and smoking habits; and they chewed their nails.[134]

Other staggering similarities emerged, many of them reported in the pages of *Science* by the respected science writer Constance Holden. Sometimes, the twins met for the first time when they showed up for the Minnesota twins study. For instance, there were Oskar and Jack. They were born to a Jewish mother and a German father in Jamaica and separated at birth. The mother took Oskar back to Germany, where he became a Catholic and a member of the Hitler Youth; Jack was raised as a Jew and wound up on an Israeli kibbutz. But when they reported to Minneapolis, they were both wearing wire-rimmed glasses and mustaches, and both sported two-pocket shirts with epaulets. The researchers discovered that they both liked spicy foods and sweet liqueurs, were absent-minded, fell asleep regularly in front of the television set, stored rubber bands on their wrists and read magazines from back to front. Both men flushed the toilet before they peed.

Even Bouchard was amazed.[135] 'I frankly expected far more differences than we found so far. I'm a psychologist, not a geneticist,' Bouchard told Holden.[134] Other studies in the US and in Denmark found the same similarities.

Important to this story is that Bouchard gave his twins IQ tests. Of all the comparisons made between the twins, IQ had the highest correlation. Many have tried to demolish the study, Kamin in particular. He says it is not serious science. 'Human genetics is an almost impossible discipline. You can't do laboratory work.'[129] Of course, Bouchard did just that.

IQ and IQ testing are still controversial and to this day no one really agrees upon what the tests measure. Meanwhile, eugenics is actually being practiced routinely – in clinics that screen for an ever-increasing number of genetic signatures such as those of Down's Syndrome, Tay–Sachs and thalassemia.

How much influence did Bill Shockley have in the IQ debate?

Virtually none.

Shockley did very little original research. He mainly did new statistical analyses of other people's data. He published a paper or two that others cited, and made some statistical contributions, but he added very little to the field. When Herrnstein and Charles Murray published *The Bell Curve*, Shockley was relegated to one mention in the foreword – much to Emmy's displeasure – where he was described as brilliant and eccentric. Jensen thought Shockley's main contribution was to distract opponents so that he and Herrnstein could get some work done.

'I have always been amazed that someone as bright as he could have contributed so little over so long a span of time,' Jensen said.[62]

—<

Shockley's urologist retired in the mid-1980s.

Shockley's prostate was enlarged, hardly unusual for a man approaching his eighth decade, but with no one to see, Shockley let a couple of years go by before finding another doctor. When he did, at the Palo Alto Medical Clinic, they confirmed an enlarged prostate and suggested it be watched. His blood test was somewhat on the high side, but the test is notorious for false positives.

In 1987, after one such test, his new doctor performed a biopsy on Shockley's prostate and found that it was cancerous. Usually, with men Shockley's age, doctors do not like to do dramatic procedures – at 77 the odds are that something else will kill you before the prostate does. But one day Shockley's leg hurt and he developed a lump on it, so he went to the clinic for X-rays.

'It had metastasized to his bones,' Emmy recalled. 'The X-rays of his bones were appalling to look at. Great big black lumps all over it.'

He began radiation treatment at the Stanford Medical Center, which made his skin hurt to touch and gave him a miserable sore throat. He complained that he couldn't drink his manhattans. He coughed often as mucus built up in his throat. Other lumps appeared, including one behind his ear. Doctors feared the cancer had reached his brain, but that was not so.

'He knew there was nothing more that they could do than what they were doing. Nobody ever talked about dying until the last.'

The doctors urged him to contact his children. He refused and told Emmy not to call either. The doctors tried to talk him into it several times, but he would not.

Surgeons removed his testicles – called an orchiectomy, a normal procedure in these cases – to cut the amount of testosterone and help ease the pain in his bones. Emmy set up a hospital bed in the family room, with a commode placed next to it. For a while, he could use the commode but then could not even get out of bed.

They did not talk about his impending death much, Emmy said. But one night, after the lump behind his ear was removed, she said, 'Sweetie, what am I going to do without you?'

'It will be hard,' he said.

'I wish I could change places with you,' she said.

She meant it. 'I wanted to die first,' she says now. 'I was a coward. It's hard on the person who is left. Very hard. Very, very hard.'

They never discussed religion or life after death. The Shockleys were agnostics and he would simply say, 'When you're dead, you're dead.'

Emmy returned to being a nurse, with the help of aides brought in by a local hospice.

In July, two of their remaining friends, a couple, called. The man had missed Shockley at the Bohemian Grove encampment and he knew he never let anything interfere with that. Not knowing Shockley was deathly ill, they invited the Shockleys to brunch.

'He's been sick,' Emmy told them, 'and he's not going to make it. He's in bed here.'

The couple came and spent several hours with Shockley, which he seemed to greatly enjoy.

Still, he would not let Emmy call his children.

'Why? I didn't ask him,' she says. 'What he said was what I did.'

He also told her not to call them when he died. 'That was his choice.... He never told me [why] and I never asked him. That was his decision to make. I didn't say, isn't this peculiar, isn't this strange. I asked him would he like me to do this. He said no, that's what he wanted. I accepted it.' She told Alison later she thought he didn't want to trouble them.

Shockley was taking Percoset by mouth to kill the intense pain. Cancer that has spread to the bones can be awful. Finally, when he couldn't swallow the Percoset, he was placed on a morphine pump. He had an in-groin catheter to help him void because the pain when he tried to urinate naturally was too much. He began to drift in and out of consciousness. He often remarked before he was ill that if he was mentally alert in the last five minutes of his life, he would think about his efforts to improve the human race though eugenics. If he even had the will to try, it would have been lost in a cloud of morphine.

Emmy sat by the bed. He sometimes would call out, and she would tell him, 'I'm here sweetie. I'm over here. Can I help?'

On the night of 11 August, a Friday, the usual aides were not on duty and a nurse named Frank came. Around midnight, Shockley's breathing changed, but it didn't register with Emmy. She went to take a nap. Two hours later, Frank woke her and told her she better come into Shockley's room.

At 2:30 in the morning of 12 August 1989, in his 80th year, Bill Shockley died. Except for Emmy, he was terribly alone.

—<

Epilogue

Alison read about her father's death in the *Washington Post*. Emmy, obeying her husband's last order, did not call her or Shockley's sons, but she did call Joan and Phil Cardon, the couple whose dinner party introduced her to Shockley. They called the *Post*. She also called Beyers, who sent out a press release that was picked up by papers around the world, which is how Bill and Dick found out.

Emmy left her husband's body on the bed until the funeral home came for it in the morning. She had it cremated and placed in an urn at the Alta Mesa Memorial Park in Palo Alto. She did not have a memorial service. It's not clear who would have come.

Alison phoned; Bill eventually came to see her. She didn't hear from Dick until much later.

Emmy left the house essentially the way it was the afternoon Shockley went to bed for the last time for more than a decade. (She finally shipped most of the files to the Stanford archives only in 1996.) His exercise machine was still in the kitchen ten years later. Shockley's office remained exactly the way he left it for years, with all his papers piled on the desk, his pencils and pens and telephone and tape recorder all ready to go. Emmy could walk by the room, and look in and imagine he was out for a bit and would be back any moment.

When I began my quest to find out about Shockley's life, Emmy kindly let me use his office. His presence in the house, especially in the office, was remarkable. At any moment, I expected him to erupt through the doorway, steel grey eyes flashing, wanting to know what the hell I was doing at his desk. Sometimes I even jumped at noises in the house.

Emmy missed him, and did every day of her life. She lived alone for 18 years in the house, becoming increasingly deaf, and eventually came under the care of two kindly Asian women. The house remained unchanged. Emmy died on the last weekend of April 2007 at the age of 94 – the exact date is unclear.

Bibliography

The vast amount of resources used in this book comes from the Shockley papers in the archives of the Stanford University Library – all 60 linear feet of them. Unless otherwise noted, the sources were taken from those archives or the library itself. Other abbreviations used in this bibliography:

WBS   William B. Shockley
MBS   May Bradford Shockley
WHS   William Shockley Sr.
JBS   Jean Bailey Shockley
ELS   Emmy Lanning Shockley
JNS   The author
SUNS   Stanford University News Service (press releases)

Part I
1.   Fredrick Seitz, 22 November 1995, JNS, Palo Alto, California.
2.   WHS, 'Diary' (1909–1910), Unpaged.
3.   WHS, 10 November 1912 (London: SWB, 1912).
4.   WHS, 'Diary' (1911–1913), Unpaged.
5.   WHS, 'Diary' (1912–1913), Unpaged.
6.   WHS, 'Diary' (1913), Unpaged.
7.   MBS, 6 May 1910 (London: SWB, 1910).
8.   WHS, 'Diary' (1914), Unpaged.
9.   WHS, 19 December 1913 (Palo Alto, CA: Putnam, A. H., 1913).
10. Shirley Thomas, 'William Shockley,' in *Men of Space, Profiles of the Leaders in Space Research, Development, and Exploration.* Radnor, PA: Chilton Book Company, 1962, pp. 170–205.
11. May's Communicative Things.
12. MBS, Diary (1918), Unpaged.
13. MBS, Diary (1925), Unpaged.
14. Alison Shockley Ianelli, 10 February 1996, JNS, Gaithersburg, MD.
15. WBS, Diary (1955).

16. Daniel J. Kevles, *The Physicists: The History of a Scientific Community in Modern America.* New York: Alfred A. Knopf, 1977.
17. Gale E. Christianson and Edwin Hubble, *Mariner of the Nebulae.* New York: Farrar, Straus and Giroux, 1995.
18. Dean Woolridge, 21 August 1976, Lillian Hoddeson.
19. WBS, 1 July 1929 (Palo Alto: MBS, 1929).
20. WBS, 10 September 1974, Lillian Hoddeson, Murray Hill, NJ.
21. Frederick Seitz, *On The Frontier: My Life In Science.* New York: American Institute of Physics, 1994.
22. WBS, 24 September 1932 (Cambridge, MA: MBS, 1932).
23. WBS, Undated (Cambridge, MA: MBS, 1932).
24. WBS, 2 November 1932 (Cambridge, MA: MBS, 1932).
25. James Brown Fisk, 24 June 1976, Lillian Hoddeson, Murray Hill, NJ.
26. WBS, 25 September 1933 (Cambridge, MA: MBS, 1933).
27. JBS, 31 October 1933 (Cambridge, MA: MBS, 1933).
28. JBS, 30 December 1933 (Cambridge, MA: MBS, 1933).
29. JBS, 26 January 1934 (Cambridge, MA: MBS, 1934).
30. JBS, 23 January 1934 (Cambridge, MA: MBS, 1934).
31. JBS, 14 February 1934 (Cambridge, MA: MBS, 1934).
32. JBS, 31 January 1934 (Cambridge, MA: MBS, 1934).
33. JBS, 28 March 1934 (Cambridge, MA: MBS, 1934).
34. JBS, 29 April 1934 (Cambridge, MA: MBS, 1934).
35. JBS, 19 May 1934 (Cambridge, MA: MBS, 1934).
36. JBS, 11 May 1934 (Cambridge, MA: MBS, 1934).
37. JBS, 7 July 1934 (Cambridge, MA: MBS, 1934).
38. JBS, 7 July 1934 (Cambridge, MA: MBS, 1934).
39. Philip M. Morse, *In at the Beginnings: A Physicist's Life.* Cambridge, MA: MIT Press, 1977.
40. Herman Feshbach and K. Uno Ingard (eds.), *In Honor of Philip M. Morse.* Cambridge, MA: MIT Press, 1969.
41. WBS, 22 March 1936 (Cambridge, MA: MBS, 1936).
42. Gibbons, 25 March 1997, JNS, Stanford, CA.

Part II

1. Alan Holden, 30 July 1974, Lillian Hoddeson, New Vernon, NJ.
2. Morgan Sparks and Betty Sparks, 20 October 1995, JNS, Albuquerque, NM.
3. JBS, 3 September 1936 (Cambridge, MA: MBS, 1936).
4. JBS, 19 December 1936 (New York: MBS, 1936).
5. WBS, 29 March 1937 (New York: MBS, 1937).
6. WBS, 11 October 1937 (New York: MBS, 1937).

7. WBS, 25 January 1937 (New York: MBS, 1937).
8. Alison Shockley Ianelli, 10 February 1996, JNS, Gaithersburg, MD.
9. WBS, 23 September 1936 (New York: MBS, 1936).
10. JBS, 28 November 1937 (New York: MBS, 1937).
11. James Brown Fisk, 24 June 1976, Lillian Hoddeson, Murray Hill, NJ.
12. John Pierce, 23 October 1995, JNS, Palo Alto, CA.
13. JBS, 11 September 1938 (New York: MBS, 1938).
14. JBS, 30 August 1937 (New York: MBS, 1937).
15. JBS, 9 September 1939 (New York: MBS, 1939).
16. R. Brattain, 24 June 1996, Interview, JNS, Monterey, CA.
17. Walter Houser Brattain, *Autobiography* (1959).
18. Walter Houser Brattain, January 1964 [no date], A. N. Holden, W. J. King, Murray Hill, NJ.
19. 'Outline of the History of Development of the Device With the Proposed Name of "Transistor"' (1948).
20. Ernest Braun and Stuart MacDonald, *Revolution in Miniature: The History and Impact of Semiconductor Electronics*. London: Cambridge University Press, 1978.
21. WBS, 10 September 1974, Lillian Hoddeson, Murray Hill, NJ.
22. Lillian H. Hoddeson, *The Discovery of the Point-Contact Transistor*. Urbana-Champaign, IL: University of California, 1981, p. 76.
23. WBS, 'The Path of the Conception of the Junction Transistor,' *Electron Devices* **23**(7) (1976): 597–620.
24. Shirley Thomas, 'William Shockley,' in *Men of Space, Profiles of the Leaders in Space Research, Development, and Exploration*. Radnor, PA: Chilton Book Company, 1962, pp. 170–205.
25. WBS, 5 October 1939 (New York: MBS, 1939).
26. WBS, 13 November 1940 (New York: MBS, 1940).
27. JBS, 5 August 1940 (estimated) (Gillette, NJ: MBS, 1940).
28. Walter Houser Brattain, 28 May 1974, Charles Weiner, Walla Walla, Washington.
29. WBS, 10 July 1940 (New York: MBS, 1940).
30 James Fisk, 'Untitled' (Bell Laboratories, 1940).
31. James B. Fisk and WBS, 'A Report to Mr. M.J. Kelly Concerning Uranium as a Source of Power' (Bell Laboratories, 1940).
32. Lyman J. Briggs, 18 February 1946 (Washington: O. E. Buckley, 1946).
33. E. P. Wigner, 19 January 1948 (Princeton, NJ: James B. Fisk, 1948).
34. JBS, 17 March 1941 (Gillette, NJ: MBS, 1941).
35. Conyers Herring, 25 January 1996, JNS, Stanford, CA.
36. JBS, 19 June 1941 (Madison, NJ: MBS, 1941).
37. Philip M. Morse, *In at the Beginnings: A Physicist's Life*. Cambridge, MA: MIT Press, 1977.

38. WBS, 'Untitled' (1942), Unpaged.
39. Section C-4 ASW Operations Research Group, NDRC, 'An Analysis of Antisubmarine Aircraft Patrolling by the First Bomber Command for the Period May, June, July, 1942' (Navy Department, 1942).
40. Section C-4 ASW Operations Research Group, NDRC, 'Statistical Research Into Antisubmarine Operations' (Navy Department, 1942).
41. Henry Stimson, 19 February 1944 (Washington: WBS, 1944).
42. JBS, 23 August 1942 (Summit, NJ: MBS, 1942).
43. WBS, 7 December 1942 (London: MBS, 1942).
44. P. M. S. Blackett, 'A Note on Certain Aspects of the Methodology of Operational Research' (Unknown, 1943).
45. WBS, 3 June 1940 (New York: MBS, 1940).
46. JBS, 31 January 1943 (Madison, NJ: MBS, 1943).
47. JBS, 19 November 1942 (Madison, NJ: MBS, 1942).
48. JBS, 15 April 1943 (Brooklyn, NY: MBS, 1943).
49. JBS, 3 May 1943 (Madison, NJ: MBS, 1943).
50. Unknown, 'General Instructions for ASWORG Members' (Columbia University, 1943).
51. JBS, 6 November 1943 (Madison, NJ: MBS, 1943).
52. WBS, 8 November 1943 (JBS, 1943).
53. WBS, 'PPI Scope Interpretation Examples and Exercises' (Army Air Forces Training Aids Division, 1944).
54. WBS, 30 October 1944 (Bombay: MBS, 1944).
55. WBS, 'The Concerns of a Non-Specialist,' in *Genetics and the Future of Man*, ed. John D. Roslansky. Amsterdam: North-Holland Publishing Co., 1966, pp. 64–105.
56. William Shockley, Personal, JNS, San Diego.
57. WBS, 13 December 1944 (Brisbane, Australia: MBS, 1944).
58. WBS, 7 April 1943 (E. L. Bowles: E. L. Bowles), Memorandum.
59. WBS, 'Untitled' (1944–45), Unpaged.
60. WBS (E. L. Bowles: E. L. Bowles), Memorandum.
61. WBS, 'Discussion of a Proposed Program on the Quantitative Aspects of Modern Warfare' (War Department, 1945).
62. WBS, 'Proposals for Increasing the Scope of Casualty Studies' (Office of the Secretary of War, 1945).
63. United States Strategic Bombing Survey, 'Summary Report, European War' (Department of Defense, 1946).
64. Cpt. Robert C. Davidson, 'Report on Radar Training Conference,' (Headquarters, Victorville Army Air Field, Victorville, CA, 1945).
65. H. H. ('Hap') Arnold, 10 December 1945 (Washington, DC: WBS, 1945).
66. WBS, 18 October 1946 (Madison, NJ: MBS, 1946).
67. Alison Shockley Ianelli, 10 February 1996, JNS, Gaithersburg, MD.

68. James A. Hijiya, *Lee de Forest and the Fatherhood of Radio*. Bethelehem, PA: Lehigh University Press, 1992.
69. Tom Lewis, *Empire of the Air: The Men Who Made Radio*. New York: Edward Burlingame Books, HarperCollins, 1991.
70. 'The Versatile Midgets,' *Time*, 11 February 1952, pp. 57–9.
71. WBS, 6 February 1945 (Washington: MBS, 1945).
72. WBS, 29 January 1946 (Murray Hill, NJ: Bowles, Edward L., 1946).
73. JBS, 4 March 1946 (Madison: WBS, 1946).
74. Richard Condit Shockley, 12 October 1996, Personal, JNS, Los Angeles.
75. Frederick Seitz, *On The Frontier: My Life In Science*. New York: American Institute of Physics, 1994.
76. Conyers Herring, 'Recollections from the Early Years of Sold-State Physics,' *Physics Today*, April 1992, pp. 26–33.
77. John Bardeen, 1 December 1977, Personal, Lillian Hoddeson.
78. Ernest Braun, 'Selected Topics from the History of Semiconductor Physics and Its Applications,' in *Out of the Crystal Maze: Chapters from the History of Solid-State Physics*, ed. Lillian Hoddeson *et al*. New York: Oxford University Press, 1992, p. 697.
79. Richard R. Nelson, 'The Link Between Science and Invention: The Case of the Transistor' (Rand Corp., 1959).
80. WBS, 'The Invention of the Transistor – "An Example of Creative-Failure Methodology,"' *Proceedings of Conference on the Public Need and the Role of the Inventor* (1974), pp. 47–88.
81. Fredrick Seitz, 22 November 1995, Personal, Palo Alto, CA.
82. WBS, 10 September 1974, Lillian Hoddeson, Murray Hill, NJ.
83. Karl Lark-Horovitz, 21 February 1948 (Purdue, Indiana: WBS, 1948).
84. 'Semiconductors,' *Business Week*, 26 March 1960, pp. 74–96.
85. 'Bell Laboratories Record' (Bell Laboratories, 1948).
86. Herbert B. Nichols, 'Mighty "Mite" Lifts Horizon of Electronics,' *Christian Science Monitor*, 2 July 1948.
87. 'The Tiny Transistor,' *Newsweek*, 6 September 1948, p. 44.
88. WBS, 9 September 1948 (Murray Hill, NJ: Malcolm Muir, 1948).
89. JBS, 22 February 1949 (Madison, NJ: MBS, 1949).
90. JBS, 14 April 1949 (Madison, NJ: MBS, 1949).
91. JBS, 27 August 1949 (Madison, NJ: JBS, 1949).
92. WBS, 20 December 1946 (Madison, NJ: MBS, 1946).
93. Vannevar Bush, 6 December 1946 (Washington: WBS, 1946).
94. Research and Development Board National Military Establishment, 'Report of Ad Hoc Committee Appointed to Study the Problem of Weapons System Evaluation' (Department of Defense, 1948).
95. JBS, 21 November 1949 (Madison, NJ: MBS, 1949).
96. JBS, 8 December 1949 (Madison, NJ: MBS, 1949).

97. Ronald Kessler, 'Absent at the Creation,' *Washington Post Magazine*, 6 April 1997, pp. 17–31.
98. Alison Shockley Ianelli, 14 April 1954 (WBS, 1954).
99. WBS, 3 August 1950 (Valouis, France: May Bradford Shockley, 1950).
100. '2 Physicists To Get Medals for Transistor,' *San Francisco Chronicle*, 13 March 1951.
101. Bell Telephone Laboratories Publication Department, 'Information Bulletin,' (Bell Telephone Laboratories, 1951).
102. Robert C. Cowan, 'Radio Tube Tested,' *Christian Science Monitor*, 16 April 1952.
103. John Rhea and Paul Plansky, ''Twas Two Days Before Christmas,' *Electronic News*, 18 December 1972, 1, 4, 51.
104. Laura and Guy Waterman, *Yankee Rock and Ice, A History of Climbing in the Northeastern United States*. Harrisburg, PA: Stackpole Books, 1993.
105. Marion Harvey Softky, Personal, JNS, Menlo Park, CA.
106. C. Hartley Berry, 13 May 1953 (Summit, NJ: WBS, 1953).
107. Janine Roger, 6 November 1953 (Rheims, France: WBS, 1953).
108. ELS, Personal, JNS, Stanford, CA.
109. WBS, 30 June 1954 (Salt Lake City, UT: ELS, 1954).
110. WBS, January 1955 (Washington, DC: ELS, 1955).
111. WBS, 22 May 1954 (Pasadena, CA: ELS, 1954).
112 ELS, 11 April 1954 (Rockville, MD: WBS, 1954).
113. ELS, 18 April 1954 (Rockville, MD: WBS, 1954).
114. 'Scientist Cites Russ Technical Training Lead,' *Los Angeles Times*, 3 February 1955.
115. Marion Harvey Softky, 6 July 1955 (Frankfurt, West Germany: WBS, 1955).
116. Marion Harvey Softky, 25 May 1955 (Frankfurt, West German: WBS, 1955).
117. Marion Harvey Softky, 6 June 1955 (Frankfurt, West Germany: WBS, 1955).
118. Marion Harvey Softky, May 1955 (Frankfurt, West Germany: WBS, 1955).
119. WBS, 16 July 1955 (Washington: MBS, 1955).
120. JBS, 5 August 1955 (Reno: MBS, 1955).
121. WBS, 17 June 1955 (Washington: MBS, 1955).
122. 'Faculty Appointments, Promotions Announced,' *Stanford Today*, 15 August 1956, p. 3.
123. Walter Brattain, 'The Saga of an Expedition to Stockholm, Sweden, December, 1956'.
124. Mary Clouthier, 6 May 1997, Personal, JNS, Stanford, CA.
125. Ellis A. Johnson, 3 November, 1950 (Washington: WBS, 1950).
126. WBS, 18 June, 1950, (London: May B. Shockley, 1950).
127. WBS, 1 June 1955 (Washington: ELS, 1955).
128. WBS, 23 June 1955 (Washington: MBS, 1955).
129. WBS, 23 June 1955 (Washington: MBS, 1955).

130. 'Noted Scientist Joins Beckman,' *Los Angeles Times*, 23 September 1955.
131. ELS, 31 July 1997, Personal, JNS, Palo Alto, CA.
132. WBS, 2 September 1956 (Garmisch-Partenkirchen, Germany: ELS, 1956).
133. MBS, Diary (1956), Unpaged.

Part III

1.   James Gibbons, 15 September 1955, Personal, JNS, Stanford, CA.
2.   Robert E. Bedingfield, 'Along the Highways and Byways of Finance,' *New York Times*, 27 November 1955.
3.   Tom Forester, *Silicon Samurai; How Japan Conquered the World's IT Industry*. Cambridge, MA: Blackwell, 1993.
4.   Everett M. Rogers and Judith K. Larsen, *Silicon Valley Fever*. New York: Basic Books, 1984.
5.   Fred Terman, 30 September 1955 (Stanford, CA: WBS, 1955).
6.   John Linvill, 15 September 1995, Personal, JNS, Stanford, CA.
7.   Annalee Saxenian, 'The Genesis of Silicon Valley,' in *Silicon Landscapes*, ed. Peter Hall and Ann Markusen. Boston: Allen & Unwin, 1985.
8.   Glenn Brown, 'Shockley planning to produce his junction transistor in area,' *Palo Alto Times*, 13 February 1956.
9.   Glenn Brown, 'Dr. Shockley enthusiastic over project,' *Palo Alto Times*, 14 February 1956.
10. Gordon Moore, 24 June 1997, Personal, JNS, Santa Clara, CA.
11. Gordon Moore, *Silicon Genesis, the Stanford Silicon Valley Project*. Tiger Productions, Videotape.
12. Fred Warshorfsky, *The Chip War*. New York: Charles Scribner's Sons, 1989.
13. WBS, 'On Research Department Productivity and Recruiting Practices' (Bell Telephone Laboratories, 1953).
14. Shirley Thomas, 'William Shockley,' in *Men of Space, Profiles of the Leaders in Space Research, Development, and Exploration*. Radnor, PA: Chilton Book Company, 1962, pp. 170–205.
15. WBS (1957).
16. ELS, Personal, JNS, Stanford, CA.
17. ELS, 31 July 1997, Personal, JNS, Palo Alto, CA.
18. WBS, '23 May–10 Jun 57 Trouble' (1957).
19. Elmer Brown, 20 August 1957 (Mountain View, CA: WBS, 1957).
20. 'Wall Street: The Yankee Tinkerers,' *Time*, 25 July 1960, 62–7.
21. WBS (1957).
22. L. N. Duryea, 28 May 1959 (Unknown: Arnold Beckman, 1959).
23. WBS (1957).
24. 'Beckman sells Shockley firm; new owners to expand company,' *Palo Alto Times*, 4 April 1960.

25. 'The Incubator,' *Forbes*, 1 May 1960.
26. Don Webster, 'Fairchild laying off about 250 employees,' *San Mateo Times*, 13 May 1960.
27. 'Nobel-Prize winner injured in accident,' *Palo Alto Times*, 24 July 1961.
28. Frederick Seitz, 30 November 1989 (London: *Nature*, 1989).
29. WBS (1961).
30. WBS (1961).
31. Eric R. Quinones, 'Bill and the billionaires,' *San Jose MercuryNews*, 29 September 1997, p. 6A.
32. 'Valley No. 1 exporter,' *San Jose Mercury News*, 30 September 1997, p. 1.
33. James Wennblom, 13 November 1997, telephone, JNS.
34. Victor Cohn, 'Will Man Survive? Yes, say 16 of 25 Nobel Scientists,' *Minneapolis Sunday Tribune*, 5 May 1963, p. 1.
35. Richard Condit Shockley, 12 October 1996, Personal, JNS, Los Angeles.
36. WBS, 22 September 1975 (Stanford, CA: Polykarp Kusch, 1975).
37. WBS, 'The Concerns of a Non-Specialist' (paper presented at the Genetics and the Future of Man, Gustavus Adolphus College, St Peter, MN, 1966), 64–105.
38. Edgar M. Carlson, 21 January 1965 (St Peter, MN: WBS, 1965).
39. John Linvill, 15 September 1995, JNS, Stanford, CA.
40. J. M. Pettit, 6 June 1963 (Stanford, CA: WBS, 1963).
41. ELS, Personal, JNS, Stanford, CA.
42. William Alden Shockley, Personal, JNS, Marina del Rey, CA.
43. 'Is Quality of U.S. Population Declining?,' *U.S. News & World Report*, 22 November 1965, 68–71.
44. WBS, 16 November 1965 (Stanford, CA: Robert Lamar, 1965).
45. Walter F. Bodmer *et al.*, October 1966 (Stanford CA: *Stanford M.D. Magazine*, 1966).
46. Rae Goodell, *The Visible Scientists*. Boston: Little, Brown, 1977.
47. WBS, October 1966 (Stanford, CA: *Stanford M.D. Magazine*, 1966).
48. Robert Lamar, in SUNS (1966).
49. NAS, 'Racial Studies: Academy States Position on Call for New Research,' (Washington, DC: NAS, 1967).
50. AP, 'Shockley urges study to determine racial mental levels,' *Palo Alto Times*, 3 September 1969.
51. Alan L. Stoskopf, 'Confronting the Forgotten History of the American Eugenics Movement,' *Facing History and Ourselves News*, Undated 1992, pp. 3–9.
52. WBS, 9 April 1965 (San Francisco, CA: J. A. Morton, 1965).
53. J. A. Morton, 12 April 1965 (Murray Hill, NJ: WBS, 1965).
54. Frank D. Leamer, 9 April 1965 (Murray Hill, NJ: WBS, 1965).
55. WBS, 17 May 1965 (Los Altos, CA: J. A. Morton, 1965).

56. WBS, 'Commonwealth Club of San Francisco,' in San Francisco (1967).
57. 'NAS Again Says No to Shockley,' *Science*, 8 May 1970, 685.
58. David Perlman, 18 January 1967 (San Francisco, CA: Editor, *San Francisco Chronicle*, 1967).
59. WBS, 29 November 1967 (Stanford, CA: Editor, *San Francisco Chronicle*, 1967).
60. Ole R. Holsti, 18 January 1967 (Stanford, CA: *Stanford Daily*, 1967).
61. Scott Moore, 'Negro Prof's Scathing Attack Shocks Shockley,' *San Jose Mercury*, 23 February 1968.
62. Arthur Jensen, 28 October 1997, telephone, JNS.
63. Christopher Jenks, 'Intelligence and Race,' *New Republic*, 13 September 1969, pp. 25–9.
64. Mark Snyderman and Stanley Rothman, *The IQ Controversy, the Media and Public Policy*. New Brunswick, NJ: Transaction Books, 1988.
65. WBS, 24 April 1969 (Stanford: Kenneth Pitzer, 1969).
66. Deirdre Carmody, 'Canceling of Controversial Talk on Negro Divides Educators,' *New York Times*, 10 May 1968.
67. 'The Ominousness of Silence,' *Bergen Record*, 13 May 1968.
68. Ray Christiansen, 'Shockley to Do Own Race Tests,' *San Francisco Examiner*, 6, May 1968.
69. WBS, 4 January 1968 (Stanford, CA: 1968).
70. John Sedgwick, 'The Mentality Bunker,' *GQ*, November 1994, pp. 228–51.
71. Grace Lichtenstein, 'Fund Backs Controversial Study on "Racial Betterment",' *New York Times*, 11 December 1977.
72. Herbert L. Packer, 8 January 1968 (Stanford, CA: WBS, 1968).
73. WBS, 15 May 1968 (Stanford, CA: J. M. Pettit, 1968).
74. J. M. Pettit, 29 May 1968 (Stanford, CA: WBS, 1968).
75. J. M. Pettit, 14 June 1967 (Stanford, CA: William R. Rambo, 1968).
76. WBS, '"Cooperation Correlation" Hypothesis For Racial Differences in Earning Power,' 29 April 1970.
77. Fredrick Seitz, 22 November 1995, Personal, Palo Alto, CA.
78. Rudy Abramson, 'Science Group Won't Drop Classified Work,' *Los Angeles Times*, 29 April 1971, p. 18.
79. Harold M. Schmeck Jr, 'Science Group Balks Study of Race and Environment,' *New York Times*, 29 April 1971.
80. 'Is Intelligence Racial?,' *Newsweek*, 10 May 1971, pp. 69–70.
81. Jerry Hirsch, 'To "Unfrock the Charlatans",' *Sage Race Relations Abstracts* No. 6, May (1981): 1–65.
82. Mary Clouthier, 6 May 1997, Personal, JNS, Stanford, CA.
83. David Dickson, 'Case for the Plaintiff,' *New Scientist*, 22 February 1973, pp. 434–6.

84. ELS, 31 July 1997, Personal, JNS, Palo Alto, CA.
85. Bob Beyers and Jeff Littleboy, in SUNS (1972).
86. Bob Beyers, in SUNS (1972).
87. Bob Beyers, in SUNS (1973).
88. Bob Beyers, in SUNS (1972).
89. Dan A. Lewis, 10 July 1972 (Stanford, CA: WBS, 1972).
90. Richard W. Lyman, 14 August 1972 (Stanford, CA: Edwin Good, 1972).
91. Jay A. Miller and Larry Sleizer, 19 September 1972 (Palo Alto, CA: Dan A. Lewis, 1972).
92. Rich Jaroslovsky, 'Harvard Forum Axes Shockley IQ Debate,' *Stanford Daily*, 19 October 1973, p. 1.
93. Bob Beyers, in SUNS (1973).
94. William F. Buckley Jr, 'Shockley at Yale,' *New York Post*, 25 April 1974.
95. Mark Singer, 'Twelve Students Suspended in "Shockley Affair",' *Yale Alumni Magazine*, June 1974.
96. William F. Buckley Jr, 'Shockley Again...,' *San Jose Mercury*, 19 June 1974.
97. ELS, 6 April 1995, JNS, Palo Alto, CA.
98. WBS, May's Communicative Things (1977).
99. Alison Shockley Ianelli, 10 February 1996, JNS, Gaithersburg, MD.
100. Lynn Closway, 'Shockley skirmish fizzles,' *Mankato Free Press*, 1 October 1975, pp. 1–2.
101. Syl Jones, 'IQ tests – A not-so-simple matter of black and white,' *Modern Medicine*, 1 February 1975, pp. 49–70.
102. Glenn Bunting, 'Shockley: Is Anybody Listening?,' *Cal Today*, 9 August 1981.
103. Bob Goligoski, 'Race-intelligence theory gets Shockley a cold shoulder,' *St. Paul Pioneer Press*, 2 October 1975.
104. Everett Caril Ladd Jr and Seymour Martin Lipset, 'Should Any Research Topics Be Off-Limits?,' *Chronicle of Higher Education*, 15 March 1976.
105. L. L. Cavalli-Sforza, 'The Genetics of Human Population,' *Scientific American*, September 1974, pp. 81–9.
106. 'Nurture, Nature, and Responsibility: Two Scientists Here Air Their Views,' *Harvard Gazette*, 16 January 1976.
107. Patrick Young, 'Is Behavior Inherited,' *The National Observer*, 16 August 1975.
108. Helen Fisher, '"Wilson," They Said, "You're All Wet!",' *New York Times*, 16 October 1994, pp. 15–16.
109. Stephen J. Gould, 'Racist Arguments and IQ,' *Natural History*, May 1974.
110. Martin Andrews, 'A Simple Question: Are IQ Tests Instruments of Class Oppression?,' *Change*, March 1982.
111. Jerry Hirsch, 15 February 1990 (London: *Nature*, 1990).

112. Richard Lacayo, 'A Theory Goes on Trial,' *Time*, 24 September 1984, 62–3.
113. Robert Beyers, 26 February 1997, JNS, Stanford, CA.
114. William J. Broad, 'A Bank for Nobel Sperm,' *Science*, March 1980.
115. Linda Yglesis, 'Sperm Banks,' *San Francisco Examiner*, 6 February 1984, p. B9.
116. Robert Lee Hotz, 'Robert Graham, Founder of Exclusive Sperm Bank, Dies,' *Los Angeles Times*, 18 February 1997, p. A-3.
117. Glenn Garelik, 'Are the Progeny Prodigies?,' *Discover*, October 1985, 45–84.
118. WBS, 5 February 1981 (Stanford, CA: Nobel Laureates, 1981).
119. Gary Diedrichs, 'William Shockley: Racist or Realist?,' *Hustler*, August 1980.
120. WBS, 3 July 1980 (Stanford, CA: Editor, *New York Times*, 1980).
121. Roger Witherspoon, *The Palo Altan* (from the *Atlanta Constitution*), 8 August 1981.
122. WBS, in Self-published (1981).
123. AP, 'Scholar calls Shockley plan similar to Nazis,' *San Francisco Examiner*, 13 September 1984.
124. William E. Schmidt, 'Trial may focus on race genetics,' *New York Times*, 6 September 1984.
125. Peter David, 'Shockley back in the limelight,' *Times Higher Education Supplement*, 21 September 1984.
126. G. Barry Golson, ed., William Shockley, Vol. 2, *The Playboy Interviews*. New York: Perigee Books, The Putnam Publishing Group, 1983.
127. WBS, 20 December 1974 (Stanford, CA: Syl Jones, 1974).
128. WBS, 'A Comment on Press Reactions to my References to "Regression"' (1980).
129. Alison Shockley Ianelli, 14 April 1954 (WBS, 1954).
130. John Linvill, 29 September 1955 (Stanford, CA: R. L. Wallace Jr, 1955).
131 James Gibbons, 25 March 1997, JNS, Stanford, CA.
132. Daniel Goleman, 'Major Personality Study Finds That Traits Are Mostly Inherited,' *New York Times Education Digest*, December 1988, pp. 89–90.
133. Stanley N. Wellborn, 'How Genes Shape Personality,' *U.S. News & World Report*, 13 April 1987, pp. 58–62.
134. Constance Holden, 'Identical Twins Reared Apart,' *Science*, 21 March 1980, p. 1323.
135. Constance Holden, 'Genes and Behavior: A Twin Legacy,' *Psychology Today*, September 1987, pp. 18–19.

Acknowledgments

Three people are mainly responsible for my doing *Broken Genius* and none of them had any idea what we were all getting in to.

The original idea for a biography of Bill Shockley came from Michael Riordan of UC Santa Cruz and the Stanford Linear Accelerator Center: physicist, author and loyal friend. When he was completing his fine history of the invention of the transistor, *Crystal Fire*, he took the story of Bill Shockley to that point. There was much he could not include and there was two-thirds of Shockley's life left unreported. He suggested to me that it would be a wonderful and challenging topic, and an important biography. So began a ten-year adventure that lasted until I was finally rescued by the second person I must acknowledge, my editor Sara Abdulla at Macmillan Science in London, the first editor to actually grasp the importance of the book and to appreciate my somewhat idiosyncratic style and approach. There is a long list of editors whose courage failed to match hers. She has been the perfect editor, and those who know me know that is not a sentence I've written often before.

Finally, this book would have been impossible without the kindness and encouragement of Emmy Shockley, Bill's widow. She desperately wanted his story told. Whether this book is what she had in mind, I don't know. She never said. She asked for no control over the content and was given none, but she was extraordinarily helpful, to the point of letting me virtually move into the house and set up shop in Bill's office. She had two safes blown open so I could get at the contents. She spent countless hours with me, and even in her 80s had a memory that was almost scary. She was never wrong in her remembrances and her recall of details was awesome. Now beginning her 90s, she has been waiting for this book even longer than I have. No man had a more loyal lover or supporter than Bill had in Emmy. No biographer had a better ally.

I must also thank the wondrous staff of Stanford's library system. I am never happier than when I'm in a library doing research (we all have our peculiarities) and Stanford's library is a joy – beautiful, serene, and full. Bill left his papers and memorabilia to the library and his archives are beyond complete. I spent mystifying months pouring through the dozens of boxes. I once told a friend the records contained everything except a laundry list and – I'm not making this up – the next day I found a laundry list. I have no idea why he saved it. I have no idea why he saved many of the things he saved, including things – as you will have seen – that he should not have saved. Anyway, cheers to Margaret Kimball and her magnificent staff, and her predecessor Roxanne Nilan, for their patience and competence in coping with the Shockley papers and with me.

A number of people were very generous with their time, either in person or on the telephone. Most especially, I am appreciative to Bill's three children, for whom these interviews were probably very difficult. They were candid and, I think, honest, and illuminating. I'm particularly grateful to Alison Shockley Ianelli, who besides giving me a great deal of her time, fetched some photos from her collection for me. Bill Shockley in Culver City and Dick Shockley in San Diego were open, kind and helpful. They are going to learn a great deal about their father that they didn't know before and might not particularly want to know now. I hope they do not find it too painful.

I'm also grateful to a number of others including (in no particular order), Jim Gibbons, Gordon Moore, James Wennblom, John Pierce, Morgan and Betty Sparks, Arthur Jensen, Bart Bernstein, Mary Clouthier, Harry Press, Conyers Herring, Robert Brattain, John Linvill, Bob Lamar, Fred Seitz, Constance Holden, and Marion Harvey Softky. Also a bow to the folks at Gustavus Adolphus College in Minnesota.

A number of colleagues also helped with time and support, including Debra Blum, Leslie Berlin, Bob Cowan, Dave Perlman, Mike Rodgers, Constance Holden, Rae Goodell and the late, great Victor Cohn who may have started all the trouble; and my hero, the late Bob Beyers, who kept it going and who also brought me to Stanford. I will be forever grateful.

If there are errors or misstatements in *Broken Genius* I am solely responsible and they are in no way the fault of the above.

January 2006

Every effort has been made to trace the copyright holders of material reproduced herein; if any have been inadvertently overlooked the publishers will be pleased to make the necessary arrangement at the first opportunity.

Illustrations

Figures 1–7, 10–12 and 17 are reproduced by permission of the Department of Special Collections, Stanford University Libraries.

Figures 8, 9 and 16 are reproduced with the kind permission of Alison Shockley Ianelli.

Figures 13, 15 and 18 are reproduced with the permission of Lucent Technology's Bell Telephone Laboratories, courtesy AIP Emilio Segre Visual Archives.

Figure 14 is reproduced with the permission of Lucent Technology's Bell Telephone Laboratories.

Figure 19 is reproduced with the permission of Intel Corp.

Figures 20–23 are reproduced with the permission of Stanford University News Service.

Index

MAX BORN

THE
BORN-EINSTEIN
LETTERS
1916-1955

Friendship, Politics and Physics
in Uncertain Times

Introduction by **Werner Heisenberg** Foreword by **Bertrand Russell**
New Preface by **Diana Buchwald and Kip Thorne**

THE BORN-EINSTEIN LETTERS
FRIENDSHIP, POLITICS AND PHYSICS IN UNCERTAIN TIMES
by Max Born and Albert Einstein
Introduction by Werner Heisenberg
Foreword by Bertrand Russell
New Preface by Diana Buchwald and Kip Thorne
MACMILLAN; ISBN: 1-4039-4496-2; £19.99/US$26.95; HARDCOVER

"An immensely readable personal account of Einstein's struggles with other physicists." *Washington Post*

"With a well-informed introductory essay by Buchwald and Thorne, the correspondence is a delight, enabling us to trace the development of the intriguing friendship between the two physicists and to read their views on the great themes of physics and politics of their time." *The Times Higher Education Supplement*

order now from www.macmillanscience.com

THE SCIENCE OF THE HITCHHIKER'S GUIDE TO THE GALAXY
by Michael Hanlon
MACMILLAN; ISBN 1–4039–4577–2; £16.99/$24.95; HARDCOVER;
ISBN 0–230–00890–9; £8.99/$14.95; PAPERBACK

"Adopting Adams' witty, punchy style, Hanlon's guide is a fun and vivid read. The science twinkles a little more than usual in such a zany setting... he tackles a wide range of cutting-edge topics with depth and authority." *Nature*

"Hanlon's book probes the possibilities inside the fiction with wit and scientist humour – not that you have to be a boffin to enjoy these ruminations, merely curious, as the late Adams himself clearly was." *The Herald*

order now from www.macmillanscience.com

CLIMATE CHANGE BEGINS AT HOME
LIFE ON THE TWO-WAY STREET OF GLOBAL WARMING
by Dave Reay
MACMILLAN; ISBN: 1–4039–4578–0 £16.99/$24.95; HARDCOVER;
ISBN 978–0230–00754–3; £8.99/$14.95; PAPERBACK

"Dave Reay has succeeded where so many scientists, academics and environ-mentalists have failed – in bringing climate change down to the level of the ordinary family. If you're not convinced about climate change, this book will change your mind. It may even change your life." **Mark Lynas**, author of *High Tide*

"How can David Reay be this wise, and still so funny? If you want to get to grips with your own CO_2 emissions – from air freighted grapes to the family run-around – this Edinburgh boffin has written a brilliant, incredibly motivating book. Read it and see." **Nicola Baird**, *Friends of the Earth*

order now from www.macmillanscience.com

Lightning Source UK Ltd.
Milton Keynes UK
UKOW01f2355281216
290925UK00001B/77/P